Perspectives of Nature
Collected Works

Collected Works
ISBN (print book): 978-1-955762-14-4
ISBN (ebook): 978-1-955762-15-1

Published by
The Shy Writer
www.theshywriter.org

Front Cover art: by Lilly Kosir
Silhouette of Brian Kosir against Northern Lights.
Taken in the spring of 2024,
at the HWY 33 overlook east of La Crosse, WI.

Rear Cover art: by Owen Kosir & Lilly Kosir
Owen taking pictures of eagles and the eagle he captured.
Taken in the spring of 2025,
at Eagle Bluff Park, in La Crescent, MN,
overlooking the city and the Mississippi river.

Perspectives of Nature
Collected Works

Scientifically Romantic and Experiential Nature Poetry

by: Paul Košir

ACKNOWLEDGEMENTS

I must gratefully acknowledge
Rodney Schroeter,
whose confidence from the beginning
in my scientifically romantic style of poetry
made this project imaginable.

I also must gratefully acknowledge
the contributions to this work
by the members of our
Wednesday Night Poets Group,
who helped me polish my good poems
into publishable works.

Finally, I must gratefully acknowledge
my wife, Lilly, and our sons, Brian and Owen,
for their help in completing and formatting this work.

From the Author

While pondering the readership of my poetry, it occurred to me that there may be some readers who would like to know just what Scientifically Romantic poetry is and how it has developed, as well as my journey in life, as they are intertwined.

I wrote probably no more than half a dozen poems as assignments in high school and enrolled in no English classes while in college. Although I had no expertise in the study or writing of Romantic or any other type of poetry, I fancied myself the originator of Scientifically Romantic poetry, which is scientific in content, but romantic in style, but not about romance. Romantic style poetry, perhaps originating in the 18th century, focused on nature and internal feelings rather than concrete notions prevalent at the start of the age of technology and scientific research. My poetry is scientifically accurate in explaining natural processes and phenomena in a way that reveals the beauty and experience of nature. It is all about nature, its beauty, and experience, not technology. My rhyming schemes, syllable counts, and rhythms also make my poetry Formalist.

"The Northern Lights" was the first poem I ever wrote after high school. I was attending the reception following the all-music service on Palm Sunday at the church in Madison designed by Frank Lloyd Wright. As I ate and drank, others left out the back door. For me, curiosity overcame civility rather quickly and I followed the burgeoning crowd outside. What drew everyone outside and away from the reception was the biggest and best display of northern lights I had ever seen. A few days later the beauty of this display inspired "The Northern Lights". That poem just poured out of me with little editing. And studying sciences, I already knew the science of auroras. Some of my poems such as "Northern Lights" popped out nearly fully formed and some, such as "Earth" took 20 or more years of rumination to develop. The story of the poem "Earth" was from a metaphorically, and literally, very deep, caving experience I had while I was attending college. I knew I should capture and write it down, which, more than 20 years later, I did.

While working as a Naturalist at Wyalusing State Park in Wisconsin, I saw how the study of science in Nature became the beauty of Nature and I wanted to catch those intricacies in the beauty of poetry. I wrote a few poems while at Wyalusing. I recall the one about the changing color of leaves "October Color". Also, a few poems on birdwatching were inspired at Wyalusing.

During the intervening years I continued to write some poems, such as the three poems I wrote while at Boy Scout camp with my two sons. I started writing several more poems while working at Aquinas High School where I taught Algebra and Science. Later while working at Chiledia and Catholic Charities I became involved in a writing group I stumbled upon. A great friend and former co-teacher, from Aquinas High school, Pete Donndelinger also started attending this writing group. In the activities of this writing group I met a publisher, Rodney Schroeter, who included me in some of his compilations. He was impressed with the scientific nature of my poetry and offered to format my work and guide me through the publishing process with a print-on-demand company. Rodney encouraged me to continue writing during and after my 1st and 2nd books. I still belong to a poetry group my friend Pete is running. And another member, Nicki Snyder, has taken over as my publisher. Their assistance and inspiration have been essential and motivating. After my 1st book I thought I had no more books within me, but 3 plus 2 more books proved I couldn't stop. Each new idea or nature concept or experience was irresistible. I write this, not to brag, but to show that that it can be done.

I was diagnosed with Parkinsons in my 50s. Increasing symptoms and incapacity of Parkinsons were complicated by my nearly 40+ year battle with diplopia (double vision), remnant of a serious accident where I was run over by a delivery truck as I bicycled in England during a study abroad term. Most of my writing has been done during my progressive diseases but that had not stopped me. But in the end it has stopped me. I am no longer able to formulate the thoughts or physically write. It is part of the cycle of life. I am entering the stage where I do not continue to write but enter the stage where I do not write but instead look back upon what I have written and accomplished. In this message from the author I am now trying to include every last bit that I have in me. I am trying to include everything, holding nothing back. Everything I write now I want to be exhaustive yet inclusive. However, I do not write this without assistance. I have tried to do so for many months. In the end, it is my assistant, my wife and companion, who is typing for me, asking questions and formatting my thoughts for me. Asking me to approve, or to clarify and expound. As in nature life is born, grows, lives, creates new life, and moves on.

In a way, these poems are my memoir. They are the nature I have studied, experienced, and loved.

--P.K. (L.K.) 2025

Introduction to Collected Works

This work includes all poems in the previous "Perspectives of Nature" volumes 1-4 & Selected Works plus 12 new poems. The previous volumes included notes, definitions, and any helpful comments on the poems facing pages for the readers convenience while this volume is a comprehensive publication of all of the poems by Paul Kosir with the poems grouped into topic categories. A few of the new or previously published poems includes notes, citations, or definitions. Also, a few of the previously published poems have been updated slightly. The versions in this volume are the preferred versions.

The sections include poems with similar topics and are not necessarily categorically precise. For example, spiders are not an insect but are put in that grouping. Also, some poems could belong in more than one group but are placed where the author felt they belonged best.

In the final segment, Stories, the poems are ordered as they were experienced in the author's life.

--L.K. 2025

To Brian,

who's always been the best son he could be.

Table of Contents

Note: New poems are denoted with an asterisk * in front of the title.

Physical Phenomena

Processes

Stories

Atmospheric Phenomena and Weather

THE NORTHERN LIGHTS

When God, in Mother Nature's guise,
sheds Her grace upon the skies,

She magnetizes ions aerial,
preparing firmament empyreal

so that each lofty atom shines
along the Earth's magnetic lines,

in pulsing, starlit choreography,
handwritten there in bright calligraphy.

While brushing hues above the air
by virtue of a solar flare,

She weaves Her light rays into tapestry,
unveiling meteoric artistry

by drawing draperies of light
across the northern polar night,

illuminating skies ethereal,
aglow with rare Aurora Boreal.

SUNRISE

The night is done,
new day begun,
first touch of Sun
provides us light.

By morning glow,
while Sun is low,
so starts the show
with lighting slight.

Each solar ray
ignites the day
and leads the way
within our sight.

Sky's reddish hue
next turns to blue,
starts life anew,
our future bright.

SUNSET

The canvas used for sunset art
is scattered blue, from light

that penetrates and permeates
at atmospheric height.

As it nears the Earth's horizon,
the Sun emits its beams

through longer path of atmosphere,
impeding short blue streams.

The palette used for sunset art
has longer orange and pink

to dab on clouds around the Sun
till colors start to sink.

First, purple colors tops of clouds,
which quickly slip to gray.

The canvas clears and people sleep
till sunrise comes next day.

LIGHT PILLARS

Ice crystals from low cirrus clouds,
suspended in the air,

with position horizontal,
cause pillars that are rare.

The crystals act as mirrors do,
reflecting nearby light,

arrangement rather vertical,
they form a shaft that's bright.

If shaft arises near the Sun
according to our view,

we say "solar" or "sun" pillars
with sunset-colored hue.

If columns other lights adorn,
like street-lamps or the Moon,

we use not solar epithet
for pillars that formed soon,

right after sunset, cold and still,
or ere the next sunrise.

But pedestal illusory,
'tis only in our eyes.

SUN DOGS

Wispy cirrus clouds of height
at times hold ice for bending light
to form a ring-like welkin rune,
circumscribing Sun or Moon.
Each halo 'round the Sun is nigh,
near quarter quarter of the sky;
its plate-shaped crystals, left and right,
hold spectral colors, focus light,
and on each flank do mock the Sun
as dazzling parhelion.

The horizontal ice white trails
show sun dogs holding out their tails,
at heel along the sides of Sol;
their brilliant beauty plays no role,
yet often dogged suns portend
precipitation will descend.

SNOW FLAKES

Nature, with Her molecules
of water vapor, pure and clear,
coats the tiny, glistening jewels
that build a fragile chandelier.

Forming crystals delicate,
embracing with hexag'nal arms
frozen shapes most intricate,
adorned with finest feathered charms.

Nature's patterns, ne'er seen twice,
Her art in bonds of atoms pent,
interlacing threads of ice,
Her tatting is magnificent.

Glass-like sculptures drift in air
and gently flurry as they fall,
nestle on the Earth to share
a blanket white of snowflakes small.

RIME ICE

Nature's silver frosting forms
in winter fog at night,

reflecting from the shrouded moon
a muted glistening light,

which pierces through the veiling cloak
to make the darkness bright.

In stillness, cooling moisture helps
the fragile rime accrue

as midnight's chill the vapors touch
to freeze the morning dew.

From icy wisps of silent air
the frozen prisms grow,

pure water from the misty clouds
shapes crystals into snow

that unseen falls upon the twigs
and feathers boughs with white.

HOAR FROST

On cold, clear nights when soft winds blow
and stars are pinned to sky,

no clouds to blanket frigid Earth,
no snowfalls keep Her dry.

Then coldness wrings the ice from air,
at times from smokestack plumes.

It leaves the icy window scrawls
In drafty sleeping rooms.

In coldness, Mother Nature frosts
all things with crystals clear

by deposition of the ice
from water vapor near.

No liquid water e'er appears
as She prepares the jewels

that decorate the surfaces
left bare till landscape cools.

EARTH'S GIFTS

Clouds

When water vapor in the air, as much as it can hold,
strikes condensation nuclei, minute and freezing cold,

the drops that form and bits of ice composing clouds are buoyed,
suspended by slight drafts of air till currents are destroyed.

If coldness follows humid days, clouds blossom down below,
evoke awed scenes from mountaintops, then through the valleys flow.

The cotton clouds of afternoons with crisp blue summer skies
stitch puffy shapes we cannot touch, but see in our minds' eyes.

Rain

The tiny drops in colder clouds touch freezing nuclei
then grow to snow or melt to rain and fall down from the sky.

Large droplets in the warmer clouds collide and coalesce,
to fall as rain from thunderclouds, as downpours, nonetheless.

The raging storms in Summer's heat from moisture held aloft
become in Autumn all-day rains of drizzle, calm and soft,

then change to flakes in blizzard snows, increasing Winter's drifts,
that melt and sprinkle rain in Spring upon Earth's growing gifts.

LIGHTNING

The turbulence of wind aloft in mighty thunderclouds
may strip from upper particles their charged electron shrouds

to leave in higher parts of clouds a charge that's positive,
at odds with lower parts of clouds with ions negative.

Potential difference in the cloud may cause electron flow;
this lightning type, by name of "sheet," can make the whole cloud glow.

If flow of charge is cloud to cloud, and seen, but never heard,
the discharge has another name, "heat lightning" is the word.

Most dangerous and damaging is lightning cloud-to-ground,
when pent-up charge in lower clouds sends leaders earthward bound.

Descending pathways step by step, these leaders act as scouts
to meet with streamers rising up along the shortest routes.

Together form returning stroke of all the static charge,
in what we see as lightning bolt that's luminously large.

In mountaintops, the bolts intense cause "elves" and "jets" and "sprites,"
but clouds that cause all lightning strikes have ice within their heights.

With energy more powerful and crack extremely loud,
some lightning strikes the other way, from ground up to the cloud.

THUNDER

A lightning strike is hotter than the Sun
and lights the sky as much, but then is done.

The air around the bolt expands so fast,
it's heard as though a shock wave from a blast.

If vertical, the lightning sounds a crack;
less upright, thunder rolls on broken track.

The time that lapses, light till slower sound,
is used to measure where a storm is found.

Five Mississippi's counted are a mile
and make one more perceptive for a while.

HAIL

When storm clouds reach the greatest height, their anvils in cold air,
strong updrafts rise through warm and moist; and downdrafts form a pair.

The atmospheric stage is set for Nature's juggling show,
its to-and-fro beyond our sight, we only see what's low.

The drops updrafted high enough will supercool and freeze;
the frozen orbs may fall to Earth, the size of smallest peas.

Some falling ice balls melt a bit then rise and freeze again,
which adds once more to bobbing spheres a glaze of ice so thin.

Quick-freezing layers capture air, make cloudy the veneer,
solidifying slowly and the coating's crystal clear.

The more that Nature juggles hail, the larger it will grow,
and faster must the updrafts be, to larger ice balls throw.

The unseen show lets curtain fall, the raining water dries,
replaced instead by hailing stones, a brief, imperiling prize.

Once danger's passed, we marvel at the fallen, juggled hail,
stones Mother Nature could not toss, but did She really fail?

WIND

The 'magic' that we see on Earth,
                to move without a touch,

is done with pressure gradients,
                on one side there is much.

For higher pressure moves the air
                to where the pressure's light,

which causes breezes from the sea,
                but from the land at night.

When cyclones march across our land,
                which takes about a week,

these pressure systems aim the winds,
                as normal, strong, or meek;

high pressure systems, clockwise out;
                low, counterclockwise in.

Our winds, in general, westerly
                without an added spin.

The howling winds in Winter chill
                then calm at night and freeze.

The gusts of March raise kites, balloons,
                while dancing papers tease.

The gently moving summer air
                evaporates to please.

With autumn currents, pollen spreads,
                to make some people sneeze.

TORNADO

RENDING ROOFS & WRECKING HOMES
DESTRUCTION LEFT BEHIND
WRENCHING HEARTS
& ROBBING REST
DESTROYING
PEACE OF
MIND
TOR
NA
D
O

RAINBOW

As sunlight from behind begins to shine on distant rain,
the drops refract returning light and color its refrain.
The water beads as prisms act to bend the light from Sun
and separate the spectral hues that once were found in one.

On back of spheres the shades reflect and exit opposite
to bring the picture to our eyes, where images are lit.
Our retinas in rear of eyes have different colored cones,
each sends a message to the mind, which blends and sets the tones.

Chromatic bands in misty arch appear at certain height
above, beside, around the point where viewer blocks the light.
Projected on cascading rain, a moving, steady screen,
as tall and wide as sunlit sky, yet picture briefly seen.

More fleeting, dimmer, rarer still, is higher second bow;
its drops reflect an extra time, inverting color row.
No two who view this lucky sight see rays that are the same,
which makes unique the portrait seen and memory in mind's frame.

JACOB'S LADDER

Rare evenings are with spirit filled,
the excess gives us light.
With beams to build our dreams at times
to make our spirits bright.
Some see the ladder, heaven-aimed,
that Jacob dreamed one night.
Still others see the Buddha Rays
through fingers almost tight.

But science notes just scattering
by particles and haze
arriving near the setting sun
and stimulating praise,
this spirit light unusual
appears and briefly stays
inviting us to grapple with
God's scientific ways.

Astronomy

Astronomy

EQUINOXES AND SOLSTICES

Ecliptic is the path of Sol, the planets, and the Moon;
Sun closest in the wintertime and farthest out in June.

The tilt of Earth makes rays direct, for heating, this is prime;
and heat so gained, for months retained, does give us summertime.

That season starts when noontime Sun is highest in the sky;
at Summer Solstice, several days, our nearby star stands high.

Each day for months, Sol does not rise as high as day before,
till daytime Sun equates the Night, which then grows more and more.

At Equinox, the dwindling Sun passes o'er Equator;
we set our clocks as darkness creeps, no longer hour later.

The shortened days still shorter turn, till nearly reaching Yule,
their heaters on, most people give, but never lumps of fuel.

On Winter Solstice comes the gift, awaited for a year;
the days are longer, bit by bit, in coldest months, some cheer.

At Vernal Equinox we plan how soon we'll feign rebirth,
so that on Summer Solstice we'll start harvest from the Earth.

METEORS

Earth's orbit intercepts the path
left by a comet's tail,

whose rock, dust, ice, and ion gas
produce a glowing trail.

When entering Earth's atmosphere,
where they encounter drag

that lights the speeding molecules,
stressed by the ones that lag.

The streaks of light ephemeral
bedazzle conscious mind,

but focusing a moment late
frustrates one as if blind.

Ubiquitous, yet personal,
radiant are the showers.

Shared glimpses of a falling star
are deeper felt than flowers.

FULL MOONS

A spirit WOLF howls from the ages of **old**,
Full Moon after Yule with its treasures and gold.

When SNOW causes **hunger**, the moon never blue,
but may lack the "full" phase, opposing the new.

New WORMs work the soil and are eaten by **crows**,
with **Lenten** snow **crusty**, sweet maple **sap** flows.

Moss PINKs highlight **sprouting grass** growing so new.
Fish easy to catch and the robins' **eggs** blue.

More FLOWERs in May ease the **corn-planting** toil;
weeds eaten by cows add nutrition to soil.

Wild STRAWBERRYs flourish in **hot** moon of June,
in Europe it's **roses** that see the same moon.

BUCKs' antlers encased in soft velvet to show
and under the **thunder**, the **hay** stacks do grow.

The ancient fish, STURGEON, is easily speared;
when **fruit**, **grain**, **hay**, and the **green corn** can be reared.

CORN moon in September is not always maize,
main grain may be **barley** that livestock don't graze.

The HUNTER sees hunted in light of the night,
first **blood** then there's movement, tries not to lose sight.

A BEAVER cuts **frosty** trees, making a dome,
with offspring for helpers, the dome becomes home.

The COLD MOON reflected in long winter nights,
this perigee moon is the greatest of sights.

ECLIPSES

Cooling gas and dust in space
did long ago accrete
to form the planets 'round the Sun
and make the Earth complete.

The Moon was formed at later time,
its orbit an ellipse,
and, when aligned with Earth and Sun,
occasions an eclipse.

Lunar

First seen is faint penumbra
of copper-colored light,
followed by Earth's silhouette,
the umbra, black as night.

Earth's shadow, cast upon full Moon,
creeps 'cross the lunar face,
extinguishing the maria,
which vanish in black space.

Solar

The unseen orb of newest Moon
eclipses solar light;
when totally obscuring Sol,
it turns the day to night.

Moon's mountains Baily's Beads create
along the crescent Sun.
Rare shadow bands of light and dark,
appear and then are done.

THE HEAVENS

Sumerians, Akkadians, four thousand years ago,
looked up and saw in pitch black skies a starlit picture show.

As Taurus, Leo, Scorpio, and Capricorn first cast,
did usher in the seasons four and told of stories past.

The cycle of their zodiac, which moved around the sky,
was passed along to ancient Greeks, with explanation bye.

So Aristotle placed the stars and planets that were known
upon concentric crystal spheres, themselves were never shown.

With epicycles, Ptolemy explained how planets moved;
for more than fourteen hundred years he could not be disproved.

Then Tycho Brahe smashed the orbs and epicycles, too.
The planets traced elliptic arcs, Johannes Kepler knew.

They moved around the Sun, not Earth, Copernicus was sure.
In Galileo's telescope were moons of Jupiter.

The telescope that Hubble dreamed was built by modern tech,
but based on bygone scientists and answered to their beck.

CONSTELLATIONS

On nature hikes, for signs of Life, we'd use our senses keen
till darkness lets our minds connect with other Life unseen.

In constellation groups of stars, parts of celestial sphere;
are named for objects, animals, and people now not here.

The constellations, 88, most commonly defined,
Zodiacal, and 48 by Ptolemy we find.

And asterisms, groups of stars, shape pictures in the skies.
We'll look not with our telescopes but only with our eyes.

One Friday in the early eve, before we gazed at stars,
saw I professor from my school, with entourage in cars.

Mythology he taught at school, the Greek and Roman tales.
He'd tell us of Odysseus, who sailed beneath Greek sails.

Thought I, he'd help me out that night,
'fore close of dimming light.

He'll spin some yarns, tell stories, too.
I'll ask him which he'd do.

I'll point out constellations then,
the ones within his ken.

He knew so much, we'd talk all night
till morning's growing light.

The only flaw in my grand dream?
Not asking him to team.

MILKY WAY

Our galaxy was named by Greeks from legendary tale,
when Hera, nursing Heracles lost milk in glowing trail.

Astronomers in ancient times, as Hera felt dismay,
looked up to hazy patch in sky and named it "Milky Way."

They saw not hundred billion lights nor spiral arms nor bars.
From Earth, can't see 'Way's blackest hole, just forests of its stars.

The nearby stars as trees are seen, some up, some down, ahead.
For distant stars, not each is seen, galactic glow' instead.

The far-off trees give edge-on view, green stripe upon the land.
In darkened skies on cloudless nights, the galaxy's a band.

While pondering black holes and Time, or tree-lines in a park,
we may begin to understand, but find we're in the dark.

Geology

Geology

Geology

MAPS

If drawn out on a paper sheet or cast upon a screen,
the data has been published there for anyone to glean.

The key unlocks the symbols that describe Earth for the keen.
They lead us from our comfort zone to many sights unseen.

The scale determines useful size, to carry and to see.
When scale is large, big things look small and smaller things look wee.

The value of an inset map, where crucial details be,
important things of smaller size, seems obvious to me.

The magic of a map is found in neither key nor scale.
The very essence of a map from head to end of tail

transposes actuality of every hill and dale
from 3D down to two and back with ne'er in mind a fail.

Geology

EARTH

Inside my Mother Earth, I feel my senses go away:
no sight, no sound, no touch, no taste; my sentience does not stay.
As I descend, the light grows dim, my sight begins to gray
till blackness strikes my eyes so deep, there's nothing where I lay.

The sounds of life above persist, but not so where I rest;
the chamber walls are silent, in their stillness I am blest.
The air beyond turns still below, within my Mother's breast;
while sheltered here, I fail to feel my limbs that once were stressed.

Through muted lips do I perceive no sweet, no salt, no sour;
the only taste is bitterness of food from my last hour.
No longer can my nose provide the fragrance of a flower,
for scent is barely issued from these walls severe and dour.

The vault that holds my earthly self is not for me a grave.
Immobile, not inanimate, I'm willing, not a slave.
This meditation, now complete, my consciousness does save;
my corporeal self, serene, now bonded, leaves the cave.

SOIL

The bedrock of an area is parent rock of soil,
produced by Nature and by Time, no touch of human toil.

Rock broken by the freeze-and-thaw of water on the land
shaped mineral ingredients, like clay and silt and sand.

By color, texture, structure can a soil be classified;
in profile and horizons, can development be spied.

Material from Life on Earth, organic in its source,
imparts the chemistry of Life, but not the living force.

Such matter and some broken rock are all soil needs to form,
then sculpted by the climate and by weather, sun or storm.

Organic matter, decomposed, is food for plants to grow,
if in a place by Nature set, or planted in a row.

Eroded soil is gone for good, take care of where we 'tread.'
Formation takes a thousand years, so soil is limited.

Protect the soil and use it well, do not the landscape hurt.
Remember that our gift from Time, our soil, is more than dirt.

ICE-AGE IMPACT

When snows outlast the summer's heat, the flakes begin to turn
from fluffy white, compressed by weight, to grainy ice called firn.

As snowball squeezed by throwing hand refreezes and compacts,
snow on a continental scale forms massive frozen tracts.

The buildup over Canada, piled high beyond belief,
above the southern Hudson Bay, was two miles in relief.

The weight of this enormous sheet forced glacial ice to spread
so snow pack in the northern states continued to be fed,

and on this front, ice left its mark with features that it formed
on landscapes north across the land, as frozen water warmed.

When climates changed and ice withdrew, the snows replaced by rains,
they inundated, sculpted dells and leveled outwash plains.

The tidal flood of melting frost, erratics in its wake,
left kettles filled with frozen bergs that warmed to flood each lake.

The water pooled atop the mass, drawn off as if by drain,
built hill-like kames of glacial drift that settled grain by grain.

Like ice cube dropped on sandy beach, each massive glacier lobe
had drift on edge and thawed to leave moraines linked round the globe.

Beneath enormous slab of ice flowed water freezing cold,
precipitating drift that helped curved eskers fill the mold.

With no one there to see the way that drumlins got their shape,
by adding or subtracting till, 'tis hid behind Time's drape.

Ice spread around a drumlin's core to make its front edge shear
then smoothed the mound of glacial till to leave a drumlin tear.

ICE CAVE

A cave in northeast Iowa, formed just like all the rest,
helped make the corner of that state karst landscape ever blest.

The water found inside the Earth became an acid weak,
dissolving limestone, dolomite at concentration peak.

Time carved the caverns found within this ordinary cave,
whose passage follows fracture lines that Mother Nature clave.

Thus, air impounded underground is insulated well,
it's steady near the average, thermometer can tell.

Yet ice cave found in Iowa entraps the coldest air,
which circulates around its rooms so cold is everywhere.

No reason found can yet explain why ice coats walls inside,
how air can always frozen feel at entrance open wide.

In largest cave from Badlands east where cavers see their breaths,
if normal cave life tried it there, they soon would meet their deaths.

We know the reasons Earth is warm and getting warmer soon,
yet no one's sure why cave is cold in latter parts of June.

Geology

CAVES

Formation

Precipitation infiltrates, fills bedrock cracks and pores,
creating rain-fed aquifers that form our water stores.

The saturated rock beneath the water table top
reacts with carbon's oxide gas, makes acid drop by drop.

The acid weak needs Father Time to help it do its task,
so ages hence the holes are carved in Mother Nature's mask.

If water in a nearby stream erodes its bed enough,
it draws the water table down and makes the landscape rough.

The limestone rock, dissolved away, leaves sinkholes in the ground;
when deeper valleys drain the rock, then caverns will be found.

Formations

The caves so formed in regions karst build artwork where they drip,
precipitating calcium in crystals at the tip.

In humid caves the moisture dries at pace of Father Time
to build and slowly decorate the sculptures most sublime.

Straws hold evaporating drops, grow downward ring by ring,
then form, if blocked, stalactite cones that from the ceiling cling.

If water trickles down the cone, then other shapes will grow:
draperies, shawls, and bacon strips are made from stone-like flow.

Stalagmites grow in counter-sense, up from the drips that fall,
and leave their lime on growing mounds for eons small to tall.

Without a drop, helictites form from moisture in a pore;
'tis delicate, fantastic art, yet fragile to its core.

Clear, gentle streams within the caves, if blocked by rimstone dams,
have pools in which cave pearls can grow, much rarer than from clams.

THE ROCK CYCLE

≈ Igneous ≈

When magma deep in Earth is warmed,
intrusive igneous is formed.
Time slowly cools the molten rock
to minerals in position lock,
but molten rock not far below
cools quickly, makes a lava flow.
Eruptions give off noxious gas,
obsidian, volcanic glass.

≈ Metamorphic ≈

Enormous pressure brought to bear
on heated rocks already there
that change to metamorphic rock
with denser lithogenic stock.
Plutonic granite turns to gneiss
with layers flat that do not slice.
Among the oldest rocks on Earth,
four billion years since geo-birth.

≈ Sedimentary ≈

Third type is rock from sediment
that's held together by 'cement.'
The sediments of sand and mud
turn shale and sandstone under flood.
As seas advanced and then withdrew,
the lime from shells reduced and grew,
so lives of shellfish can be read
in size of every limestone bed.

EARTH'S GEMS

Geodes

In secret, Nature lays her eggs, rocks lacking artistry;
no decorations on their shells of quartz chalcedony.

Instead, their shells, when broken, show the mineral gifts inside,
from solutions saturated, there grew solutes that since dried.

A crystal palace lies within each dingy ovoid stone;
amid the gems, a seat unseen for Mother Nature's throne.

Each floor and wall with jewels bedecked, on ceilings, chandeliers.
The other geodes, still intact, unopened through the years.

Agates

As gases of the early Earth escaped its lava flows,
then Time made cooling bubble frames and fashioned studios

with outer walls of moganite, which looked like normal stones,
but inside, workshops stocked with gel-like silica for tones

to paint small frescoes on the walls at nucleation points,
adhering there with micro-crystal curvy fiber joints.

More coats, more coats till Time is done; the agate filled, unique.
When hewn, it shows the banding that the agate-hunters seek.

SANDSTONE BEAUTY

Four hundred million years ago,
the Ordovician Sea

lashed the shoreline, pummeled sand,
without impunity.

Through millions more, sand turned to stone
with iron showing hints

of many oxides colorful,
red, yellow, rusty tints.

These shades are hid in many spots,
but still considered rare,

in glens and dells and overhangs,
wherever sandstone's bare.

In Winter and the early Spring,
the overhangs are best,

endowed with frozen waterfalls,
beginning at the crest.

The frozen jewelry worn by Earth
has beauty seasonal,

but deeper colors of the Earth
are less ephemeral.

Geology

PLATE TECTONICS

A suit of armor plates the Earth with pieces of the crust
that fit together puzzle-like, into their places thrust

by hotter mantle down below that rises up and slides,
so plates tectonic pass on edge, build tension 'long the sides.

Plates also move subductively, too slow for us to see,
as lighter continental crust slides over crust of sea;

yet smashing plates will mountains build, if density the same.
There's spreading in Atlantic Ridge was Harry Hess's claim.

Volcanic action, earthquakes, too, are found near boundary,
where plates colliding drive the course of core geology.

The movement of tectonic plates, or Continental Drift,
explained by Alfred Wegener, caused scientific rift.

GEYSERS

Like a tempest in a teapot,
a geyser makes its show

if two outlets reach the surface,
both heated from below

by magma underneath the Earth
which makes the water grow.

When hot enough, the water boils,
steam rises in long spout

till it releases in a jet
and all the steam is out.

The water, just erupted forth,
fills up the 'room' and then

is heated up another time,
boils over once again.

A geyser's faithful to its blasts,
you'll always know just when.

Physical Phenomena

Physical Phenomena

Physical Phenomena

HEAT

Before the nineteenth century, "caloric" carried heat
in scientific papers and while walking down the street.

This liquid explanation of how warm things got warm
became replaced by energy, kinetic in its form.

Kinetic heat is passed along in one way out of three:
conduction, radiation, and convection we can see.

Conduction must by contact spread from warmer to the cool,
warm things, when felt, start losing heat; this state can sometimes fool.

Convective heat by currents moves, in rooms and Gulf and core.
Heat rises to the ceiling, so it's cooler near the floor.

The Gulf Stream carries water to the European states,
which makes them warm for folks to live and betters, then, their fates.

Currents in its outer core cause Earth's magnetic field
that saves us from Sun's cosmic rays, without it, fate is sealed.

By radiation, heat is passed to objects far away,
thus Ultraviolet energy is carried by Sun's ray

then through transparent barriers, including greenhouse glass,
and CO_2 or methane or some other greenhouse gas.

These U-V rays are turned to heat that's trapped below the sky
to warm the Earth and make some think disaster's coming nigh.

CAMPFIRE

The air is crisp, the tinder dried,
the fuel is set, combustion tried.

Tinder for a camping fire ignites imagination,
ere the match is even struck burns bright anticipation.

The twigs arranged in kindling pyre
like dominoes will fall to fire.

Puffing glowing sparks to flames prevents the fire from waning;
the heat-dry-fuel-and-oxygen reaction starts sustaining.

More wood creates a bigger blaze
with flickered light to hold our gaze.

Though dancing flames of ion gas seem not to be substantial,
this element from ancient times still is consequential.

Our fire, less stoked, burns down to coals
to meet marshmallow-roasting goals.

Its glowing embers in our eyes connect us with the Past,
when roasting-boiling-warming fires for living had to last.

BURN

From planning the scorch to lighting the torch,
the burning goal is weed control
to help the fate of Nature's state.

With flame for breath, new life from death;
new green from black, when plants grow back.

Invasive plants lose growing chance
when combustive fuel is management tool;
weeds' edge deprive, so natives thrive.

The glowing light of plants at night
to ash will turn in evening burn.

Smoke from the blaze adds soot to the haze,
charred logs are smelled when flares are quelled,
then dowsing tames the crackling flames.

Smoldering birch in morning search
means dwindling heat, combustion complete.

The fire done, the battle won.

SOUND

Sound travels through a medium, most oftentimes the air,
which, when disturbed, moves molecules to make them dense then rare.

Compressions and releases cause the frequency, or pitch,
the volume, and the quality that makes a noise sound rich,

the beauty made by instrument, or choir or a bird,
the melodies we sometime sing, or other music heard.

But music's not the only good in sounds that we all share,
communication, spoken, sung, shows others that we care.

All animals communicate. What sets us, then, apart?
Ideas that are new to us and poetry that's art.

A USEFUL FORCE

On surface, it keeps boulders in their place,
in elevation of the earthen face.

Environments of humans are the frame
that, held by forces, will remain the same.

Not that which keeps Earth's orbit 'round the Sun
with day-and-night rotation being spun

at perfect distance for the Life on Earth
to give its outer layer living girth.

When we, like early humans, leave our caves
by simple trail or footpath that one paves,

if in an auto or on back of horse,
our transportation needs this useful force.

With varied movements made from here to there,
perceptible, at best, or on a tear,

machines in use a hundred times a day,
a wheel, a rope, and brushing dirt away,

for harvesting and cooking all our food,
by force, we eat and drink what we have brewed.

With force, the words that humans ever spake,
on paper, now, or digital we make.

More than just opposition, friction drives
the most important actions of our lives.

FLYING

Aerodynamic shape of wings
helps birds stay in the air

while soaring or in powered flight,
wings working as a pair.

The secondaries of a bird
give wings their shape and lift,

while thrust comes from the primaries
to make birds' flying swift.

The way in which birds ply their wings
determines path of flight.

How much, how hard their pinions move
affects the flying height.

Bald eagles can do barrel rolls;
Ospreys dive feet first;

Hummers aim for nectar tubes,
to quench their nagging thirst;

Kingfishers plunge from branch for fish;
Crows 'row' their wings to fly;

male Woodcocks zig-zag up and up,
then tumble from the sky.

WATER

Unique among the molecules of compounds world around,
in liquid, gas, and solid states can water e'er be found.

All living things need H_2O, the chemical of life,
to bathe reactions aqueous, assuaging thirsting strife.

Yet Life itself depends on traits of water physical,
to help create and guide the surge of life force mystical.

Fresh water moves in lower plants by capillarity,
but vascular in higher plants, a flower or a tree.

Adhesion and cohesion move fresh water through each plant,
to transport Life's elixir to the places where it's scant.

And water shaped the faunal life that dwelt in ancient bays,
archaic fish developed gills, employing water's ways.

To fish that live in waters fresh, dissolved O_2 gives breath,
the floating ice and flow below allow for cheating death.

Cold water holds more oxygen, the life-sustaining gas,
so deeper in a frozen lake is where you'll find the bass.

SWIMMING

The ducks that dabble swim on top, the divers down below.
With bones that are not hollow, though, still deeper can loons go.

A coot will dive then pop back up, but grebes are birds that sink.
They force out the unwanted air and make their air sacs shrink.

Auks dive for fish and feed on squid, but mostly they eat krill.
Antarctic penguins dive to stay away from 'whales' that kill.

Whales with beaks and elephant seals dive deepest of them all
with packed-in blood cells, shunted flow, they make a complex call.

The most-adapted animals to life beneath the wet,
can spend it most completely in an underwater set.

In water, fish do everything, including breathing air,
extracting from it oxygen, dissolved and rather rare.

With their gills attached to arches, which look like slotted 'cheeks,'
the fish use countercurrent flow for oxygen from creeks.

In Dallol, adaptations fail in habitats most cruel;
none live in hyper-saline acid geo-thermal pool.

THREE WORLDS

Through Escher's eyes we Nature see
when we at lake count worlds at three.

With one perspective for each view,
each lake in front of us is new,

for closer to the surface seems
fish image on bent sunlight beams

and tree leaves that on surface rest
look true on every wavelet crest.

The 3-D trees before our eyes
transform to 2-dimension guise

and stretch out over water's face
to Nature and Her tree-line trace.

Three worlds are found in Nature's light,
beyond, below, and water-height.

The images in mind we make
are archetypes for every lake.

LIGHT

To the looking glass and back, describes the path of light,
reflected from its silvered rear, that's polished clean and bright.
The image seen is regular, transposed from left to right.
What else has turned to wonderland in image of our sight?

By moving to another glass, we change reflected view,
the mirror now not planar, so its image is askew,
which alters our perspective so it's something wholly new,
removing the obliquities that quite unnoticed grew.

But through the looking lens and on, will cause a bended ray.
An arrow 'hind a water glass will point the other way.
The shape of lens, convex, concave, decides how it will stray
from angle of its incidence, where it will never stay.

With mirrors and a lens inside, he built a viewing tool
to help his students learn at night in most impressive school.
As Newton worked, invented math, new physics, as a rule,
he looked at planets' orbits while he sat upon on his stool.

Through the prism and beyond, is where white light is split,
revealing spectrum locked within, six colors to be lit.
The rainbow's seen by other means, in water spray, oil bit.
Light's nature is duality, part wavy and part whit.

SUN

Within the furnace at its core,
our Sun does elemental chore

by fusing hydrogen to form
new helium in plasma storm.

Exploding forge will atoms bake,
to carbon and some others make.

Most heavy elements arise
with supernova stars' demise.

But not our Sun, its greatest grace
is radiation sent through space,

that powers weather, growth of plants,
affecting fauna's living chance,

providing and withdrawing Life,
the touch of Sun on Earth is rife.

In nightly dark, there is no Sol,
yet everywhere is solar role.

COLOR BLIND

With man-made hues, I'm color-blind; I do not know the tints
of fabrics in the fashion world or tinctures found in chintz.

I know not French nor heraldry, so not the color vert.
What I see best are Nature's tones and different shades of dirt.

Some tinges that I do not know are taupe and puce and dun;
they're artificial stylish dyes, made indoors, not in sun.

Sienna brown, I never know, if yellowish or red.
Cerise is red that's light and clear whenever it is said.

Not so for "lake," a word for red that means deep water blue,
or "madder," yellow-flowered plant, that's also reddish hue.

And "fallow" means a resting field, but also yellow shade;
I've seen this color many times, as fields begin to fade.

Cerulean is blue I know, a warbler streaked on back.
A bunting male looks indigo yet pigment only black.

I see the light, reflected blue; it always makes me smile;
not color-blind, I see what's hid – the bunting blue espial.

OCTOBER COLOR

"Nature's first green is gold," writes Frost,
"Her hardest hue to hold".

For Summer hides the golden hue,
but Autumn makes it bold.

In summer leafage, Life is green
till shorter days turn chill.

Then chlorophyll gives up the ghost,
revealing xanthophyll.

The red that's seen in autumn leaves
was not before produced,

nor always shows the scarlet hue
when green becomes reduced.

As sugar forms and thriving wanes,
vermilion shades are built.

Bright sun, with its opponent cold,
the leaves do crimson gilt.

NATURE'S CLOCKS

Radioactive rocks are placed in special hourglass,
thence change away at even rate to see the ages pass.
These elements take certain time to change one half away,
same length of Time, new half will change while other half will stay.
In using such an hourglass to time the age of rock,
one finds the age of nearby stones as sure as if by clock.
The older rocks are lower down, the younger higher still.
Time stacks things not from top to base, no matter how He will.

Uranium is often used to find the age of rocks,
but diff 'rent elements show well the lesser ticks and tocks.
For objects made in human times, the dating that works best
is carbon for organic things, a basket or a vest.
Most items found from ancient times, were made to show respect,
yet objects of prehistory survival did affect.

Processes

Processes

LEARNING NATURE

How happy is the learner who, in Time, becomes aware
that forms of Life, while similar, if common or if rare,

can be distinguished from the rest, to show diversity,
as in the ferns, the clams, and frogs and small arrays we see.

This revelation sparks the mind to know some varied types
but not enough to master groups with many diff'rent stripes.

The study of the larger sets, like fishes, birds, and trees,
is slow at first, till tipping point, then learning has more ease.

Beyond the tipping point, some know the Life that can be found,
if over, under, on the Earth, it serves as common ground.

Enormous groups still harder seem, the insects and the plants;
they're often left to experts and their sub-specific rants.

THE GRAMMAR OF NATURE

Nature does not live in the subjunctive,
conditions never contrary to facts,

expressing zero wishes and desires,
responding not with language, but with acts.

Inherent message is imperative,
the unrelenting drive for more to live,

keep going without question or a doubt,
with object of the actions still to give.

Considering no possibilities
that predicate outside the present tense,

lacking grammar for temporal thinking,
conceiving of the Future makes no sense.

Life is indicated, not suggested,
by Nature's normal ventures, not the rare,

predicting not the movements yet unseen,
She stumbles not on step that isn't there.

"ZOO"

There's never "zoo" heard in the word
that some pronounce "zo-ology."
They rhyme the "zo" with show then know
'bout lives of critters that we see.

If first two 'o's sound as in "zoo"
there's not an 'o' for "-ology,"
yet if the first 'o' rhymes with 'do,'
there's plenty for "zo-ology".

To view the life we cannot see,
try micro-(sized) biology.
To learn about the honey bee,
the path is entomology.

For those who catch fish in the sea,
their course is ichthyology.
For those who wish to make some brie,
first they should know mammalogy.

But if you want to plant a tree
for Arbor Day dendrology;
close down the zoo and you'll be free
to study plants with botany.

FLOWER ADAPTATIONS

Oh my darling, Oh my darling,
Oh my darling, columbine,

You are red with yellow insides
and the hummers drink your wine.

From the spurs on backs of flowers,
they sip nectar oh so fine.

Pollination is the mission
they accomplish as they dine...

Oh my jewelweed, Oh my jewelweed,
Oh my jewelweed, touch-me-not,

Flowers look like bee garages,
rub off pollen on the spot...

Oh my milkweed, Oh my milkweed,
Oh my milkweed, monarch food,

Bees step into flower buckets,
to their feet is pollen glued...

Oh my orchids, Oh my orchids,
Oh my orchids, they take time,

For the sights of reproduction
and the smells you also mime....

PHOTOSYNTHESIS

For 90 plus three million miles,
Sun's light beams traveled straight,

continued on to chlorophyll,
uncommon photon fate.

With light reactions' energy,
new oxygen in air,

the molecules in cycle Krebs
increased electron share,

producing things like ATP,
near thirty-eight the gain.

This done in respiration with
electron transport chain.

Next step is Calvin cycle dark
to carbohydrates build

and balance out the charges with
electron spaces filled.

The quantum, once a part of sun,
ejected from its place,

became a part of living plant,
"sun" flower sent through space.

WARMTH

In Autumn, shorter, cooler days
stir changes in the eating ways

of mammals who must often eat
to warm with metabolic heat.

While some continue on this road,
the finding-food-when-hungry mode,

a never-ending way of life
of feasting mixed with starving strife,

some species burrow underground
and wait for Spring to come around.

Some of the bats and many birds,
a butterfly and hoof-ed herds

migrate to find the better climes
to spend their hungry winter times.

As mammals, humans do the same.
As spirits, humans need a flame,

which friends and family help to light
to keep away the cold at night.

Some people hibernate or fly
to keep their bodies warm and dry,

but miss the visits, kith and kin,
the chance to feel, to talk and grin.

Yes, Autumn is the time of year
to build up warmth and gather cheer.

WINTER

Winter beginning	Arctic air winning
Noon sun sinking	Day length shrinking
Night air chilling	Hoarfrost killing
Bucks end their rutting	Burrow doors shutting
Sun not warming	Blizzard forming
Clouds stalling	Snow falling
Wind blowing	Drifts growing
Mammals hiding	Raptors biding
Thaw only teasing	Water still freezing
Night skies clearing	Cold wave nearing
Wind chill gripping	Frostbite nipping
Temperature dropping	Bird feeders hopping
Arctic air losing	Bears finish snoozing
Days longer	Sun stronger
Ice going	Sap flowing
Cold diminishing	Winter finishing

FEBRUARY

The shortest month?
Hah! That's a laugh!
Each frozen day
seems one, plus half.

FEBRUARY, TOO

In February, lips turn blue,
a month too short for Moon to do.

SYRUP

While frosty morns are warmed by day
and sunny days at night turn cold,
the life of trees begins to flow
and maples pump their liquid gold.

A year before, their leaves did catch
the strength of Sun, the breath of air
to fuel the trees' life-giving sap,
stored underground for Winter bare.

Descended through the phloem tubes
in Summer's growth and Autumn's fall,
life rested Winter under snow,
rising back through xylem tall.

In early Spring, when frozen trees
begin to thaw in mildest heat,
some tap the flowing life dilute,
distilling from it something sweet.

MARCH THAW

I heard it one day
from the shore of the lake,

while my eyes were asleep
and my ears were awake.

I heard something happen,
so I know it was true,

and yet oh so quiet,
like the sound of the dew.

Then when straining my ears,
I could hear the sound more;

'twas a thunderous crash
and a deafening roar.

So it couldn't be seen
and it couldn't be felt,

when I harkened to Spring,
I heard the ice melt.

SPRING

Each Spring, the Earth knows not the day nor date,
for Sun does not the numbered pages turn.
Must we for warmer weather have to wait?
Or has the daystar long enough to burn?

Ere sprinkles help the greening plants to bloom,
The warm at night allows the same to grow
while dainty buds detract from raining gloom
till pregnant streams with water overflow.

The minstrels sing an early daybreak song
but unresponsive maidens send no love.
To find a mate has taken far too long
with plaintive song of faithful mourning dove.

Cast out the thoughts of matters that you dread.
A match appears with sunny days ahead.

YOUNG

Born or hatched with opened eyes,
precocial young are ready;

they don't need parental help,
first steps and stance unsteady.

Some enter Life with eyes closed tight,
young altricial need more care,

all things 'round are new to them,
they begin with bodies bare.

Ungulates, the hoof-ed beasts,
in rearing are precocial.

So, too, are hares, yet rabbits, no,
they're very much altricial.

Is the womb where young develop,
as precocial lone or twin?

Or after birth, altricial,
with the rest of litter kin?

Precocial or altricial?
Which describes the human young?

Precocial by their single birth,
yet altricial lullabies sung.

PARENTING

Our children die a thousand deaths,
but other mammals, one,
for parents yearn for bygone times,
ere childhood days are done.

We watch our children grow and learn,
are proud but saddened, too;
once curious, naive, and dear;
the loss of these we rue.

We teach them lessons, guide their lives,
which starts a wistful flood
of thoughts nostalgic that we feel
about our flesh and blood.

These memories remind us that
our children's youth must end,
to be replaced by grown-up souls,
adults we need not tend.

But mammals never mark the dates
their offspring grow, mature.
Their young live life until they die,
a mass of flesh and fur.

LIFE

A beating heart begins from start,
the mind unstraps with missing gaps
that parents fill with unused frill
as children wait to see their fate.

While Nature's led by Father Time,
He lets Her live and learn,
but when they part their ended ways,
no Future to discern.

TIME

The thread of Life rolls out in Spring
to near infinity,

and ties together living things
in Time's divinity.

Our Summer tablet holds a list, a simple registry
of tasks and fancies for our lives that we in mind foresee.

We turn the pages, ledger days against imagined Time,
our Summer stolen by a count not done by final chime.

At Autumn's feast on Winter's eve,
Time cools in hibernation,

while other patrons quickly eat,
Time warms in their migration.

How long, it seems, the Winter lasts
as marked by Winter Wren,

count seven seconds for each song
that's absent from his glen.

USING TIME

The never-ending currency
of Time well-spent today

will ne'er be lost, but always found,
whatever come what may.

Struck only once by hands of Time,
but minted every day,

and crafted into food and drink,
exchanging weigh for whey.

Is ever said the gift so fine,
wrought by the other hand?

Whether made by thought or craft,
ought other be less grand?

Expense of Time cannot be judged
or rated coins in hand.

But choices that at last we make
based solely on the Land.

LIFE'S HISTORY

The hardest things in Life are ne'er forgotten,
they're set for the ages and written in stone:

the strength that suffuses a primitive shell,
the framework supporting Life cut to the bone

are clues to the mysteries unfolded in Life,
posterity left in impressions of yore,

punctuations engraved in layers of rock,
a record of Life as it's lived at its core.

Yet daily existence and struggles in life,
actions and instincts in Nature's creation,

learning in offspring and tending by parents
occur without trace in sedimentation.

Earth publishes not what living things utter,
their unnoted thoughts are forgotten and dead.

The soft parts of Life are all in the Present,
imbued with their genes for the Future ahead.

PUNCTUATED EQUILIBRIA IN LIFE

The change in species past and yet to come
is punctuated equilibrium,
should it occur so relatively fast,
in generations, not in ages vast,
caused by mutations beneficial,
not natural selection gradual,
to cause resulting incremental steps
amid so many stable, lasting reps.

Another case of this scenario
is how relationships of humans grow.
At home most children learn necessities
of eating, drinking, dressing, saying "please."
Next lessons: reading, writing, 'rithmetic,
learned in the home or at a school of brick,
curricula for teaching many sorts:
in music, acting, science, playing sports.

Among the punctuations that we face
are those in which we have to change our place.
As new adults, we leave our parents' home
and to a place of work or study roam.
We look for equilibrium in life,
perchance in role of husband or a wife
then children cause mutations in the plan
as sweetest punctuation in the clan.

As they adapt to changes in their youth
by using wisdom that we've left as truth,
while making thousand choices for their days
we see their childhood full of "yeas" and "nays."
When children leave, the quiet makes us sad,
there's no one left to call us Mom or Dad.
With children gone, there's nothing left to do,
but learn an art or craft or something new.

A HUNT

The Past is our remembrances and they become our lore,
a chance to live a second time or more or more and more.

Survival's based on memory and aptness for the kill.
The struggle is primordial, the battle rages still.

Few second chances in a hunt for animals to try.
Who flees the first, if safely done, has one less chanced to die.

No winner in the contest while combatants both alive.
No loser if, in battle fierce, contestant does revive.

In all-out war, on either side, there's little left to give.
The celebration at the bell, is more of Life to live.

They shake no hands at end of hunt in game called Life they play.
If conflict draws, they scamper off; each goes a separate way.

EROSION

With a steady, infinitesimal gait,
Time does all His work at a slow, even rate.
The persistence of Time will never abate,
even when frozen, will the surface ablate.
Time-honored erosion, need not mediate,
the work is unending, to disintegrate.

Time sharpens His tools for eroding the Land,
but tools in His kit are not built for the hand.
They have not a case nor an imprint nor brand.
With wind and some water, He beats earth to sand.
In two blinks of His eye, He dug Canyon Grand,
honed parabola Arches in rusty red band.

Habitats

Habitats

PRAIRIE

Sedges have edges and rushes are round;
but grasses are hollow from top to the ground.

First two like wetness, but grasses not so;
they're native to Plains states, where winds often blow,

growing in grasslands called prairies by name;
the rich soils beneath them made landscapes to tame.

Prairies remain on the lands undisturbed,
but only in places where plowing was curbed.

Flames on the prairie, ignited with ease,
burn all the dead grasses and smallest of trees,

properly managed and set through the years,
they guided succession, in stages, or seres.

Grassland savannahs have trees of great girth,
with deep-growing roots that help anchor the Earth.

Climax of plant life may not be the trees,
but grasses and flowers in warm Summer breeze.

THE TENSION ZONE

The State, Wisconsin, has a band, dividing south from north,
that Minnesota, Michigan, its neighbors, carry forth.

It shows where northern species grow in normal climate times
and marks the reach of southern plants in what are northern climes.

The range of more than hundred plants compiled to form the lines
on maps denoting Tension Zone; in Nature, shown by pines.

In region north of Tension Zone, some snows precede harsh cold,
whereas, in south, there's frozen ground ere bragging snows are told.

The climate of the Tension Zone controls the types of plants,
which then affect the faunal life and mammals seen by chance.

Above the line live porcupines, the star-nosed mole, and bears,
least chipmunks, woodland jumping mice, the lynx, and snowshoe hares.

To south, you'll find white-footed mice, opossum, eastern moles,
the ground squirrel marked with thirteen lines, and poorly-named pine voles.

The pine vole rarely enters pines, knows naught of Tension Zone.
It feels no tension, stress, or strain, and keeps a hungry tone.

DRIFTLESS AREA

A dry plateau in seas of ice
was spared from glacial action twice

and twice again evaded weight
of glaciers huge, with impact great.

The isolation of these lands
let rare, endangered species stand.

No drift or till ice carried in,
yet wind-blown loess again, again.

Where glaciers crept, they filled each cave,
but landscapes karst did Driftless save.

With sediments from river laid
and ice-age features never made,

the force that shaped the Driftless Zone?
Erosion carved the Earth, alone.

The valleys cleft by rill and stream,
adjoined above by ridge-top seam,

intersect as twig to limb,
dendritic flow to sea from rim.

This water drain impounds no lake,
but floodplains flat create a brake

in wetlands 'long the river shore,
where glacial outwash formed a floor.

A host of bluffs and relict pines,
man's contour farms, historic mines

are found throughout the Driftless Lands,
some made by God, some, human hands.

DESERT PAUSE

While looking right as far as I could see,
I saw no living thing except for me.

On left I watched the west wind brush the land,
with nothing on the surface save for sand.

Behind, I left a trail of shifting grains,
made desolate by lack of living rains.

Before, stood dune and yet another dune,
bedazzling and stark against the noon.

Oppressive heat on every sandy knoll,
relenting not on path to final goal.

The searing Sun intensely at my heels,
I sifted sand with spokes of pedaled wheels.

MOUNTAINS

Our
train was late,
but all was good,
we saw three moose on road,
a cow, two calves that pranced
around; their playful nature showed.
We pitched our tents in mountain
dark and slept in peace of night.
Then found we'd camped near mountaintop.
Surprise in morning light. We later saw the bighorn sheep,
a mountain goat, and bear. The mountains are for those in life
who dare to care and share. Now, those who make the mountains prime,
and also take their time, can find a place to mount a hike to see sights most sublime.

On
diff'rent trip,
with backpacks tried,
We'd hike up to the pass.
Three times I slid from slanted
tent and slept on snow, not grass.
Our scouts hiked up toward mountain pass
to find the snow neck-deep. We always knew
our plans might change, descending mountain steep.

SPRINGS

Water seeping
Bedrock weeping

The rain sinks in until it's blocked,
within perched water-table locked.

Water dripping
Pebbles slipping

Erosion, some by gravity,
while water carves a cavity.

Water flowing
Streamlet growing

Ahead of movement ever down,
the soil of Life, first black, then brown.

Water spilling
Valley filling

Then Time and Nature form the bed
of life-sustaining watershed.

Water trapped
Artesian tapped

To drive a pipe is only thing
that's needed to uncork this spring.

Water pressed
Landscape blessed

When aquifer is on a slope,
clear water rises, as does hope.

RIVER

As River passes 'fore our
eyes, in-finite water that we see
will grow in its apparent size
to hold the next infinity.

So many times is water drawn
from where it comes to where it goes,
replenishing from dawn till dawn,
the water of the river flows,

flows over ev'ry substrate stone,
each form of Life on riverbed,
and those that river's current meet,
where freely-swimming fish are fed.

If flotsam, jetsam do abound,
infinity of here declines
where everlasting now is found
we must build ecologic shrines.

BOGS

If drainage of a pond or lake is blocked by native clog
for thousand years, a wetland forms, a spongy "quaking bog."

The vegetation that can live within this acid pool
will grow in mats a meter thick, unsteady stepping stool.

A bog is poor in nutrients, most native plants need more,
but not bog's insect-eating plants, which have enough to store.

Decaying matter in a bog is starting point for coal.
Absorbent peat moss, found in bogs, provides free flood control.

The peat moss in the garden shops combats aridity.
When grown in bogs, cranberries have a mild acidity.

Amend the garden, keep it moist, red berry juice to sup.
Next time it feels that you're bogged down,
think rather, you're bogged up.

MARSH

A bur-reed marsh is mostly there,
where cat-tail plants are very rare.

While threatened species fly around,
invasive species grow from ground.

Marsh dried a Mississippi flood
to make it Mississippi mud

in the year, Two Thousand One
and '50s, '60s, still not done.

Teachers use for education,
others find their recreation,

a place where scholars do research,
and spirits treat it as a church.

Blackbird, egret, dragonfly,
a wood duck painted on the sky,

fishes, muskrats swimming past,
gosling follows mama last,

swallow, warbler, leopard frog,
turtles sunning on a log.

All these things seen in "The Marsh"
where Summer's gentle, Winter's harsh.

TIDE POOLS

In predawn light, ere curtains rose,
extreme environment we chose.
We found our seats to watch the shows
our spotlights shone, the action froze.

Tide leaves some pools, more stages grow,
echinoderms where pools have flow.
At high tide, players live below,
but cling to rock when tide is low.

Exposed to sun, no chance to flee
predaceous birds, salinity.
Another stage, diversity
of colors, shapes, and pageantry.

Dark purple, pink, and yellow, green,
the strangest forms we'd ever seen,
a starfish odd, legs seventeen,
it takes a bow to end the scene.

A hermit crab has changed its shell
for matinee to play as well.
We watched more shows before the bell,
then waters rose and curtains fell.

OCEANS

Most never lose the sight of land,
and still we say we love the sea.

Experience is beach of sand,
beyond the surf is absentee.

Our love of sea vicarious,
so what we learn becomes heartfelt.

The oceans' fates precarious
whenever ice sheets warm and melt.

But even with an airplane view,
can't fathom its immensity.

The circulation 'round, anew,
is caused by water's density.

Cold, salty water, weighing more,
submerges with Atlantic North,

moves downward on the ocean floor
to rise with Gulf Stream flowing forth.

In the ocean named "Pacific,"
where water forms gigantic gyre,

refuse has become horrific
in garbage patch of plastics dire.

Birds

Birds

Birds

WARBLERS

They never stop moving, I can't get a good look;
when they finally sit perched, it's not like in the book.

All seem to be yellow and they look just the same,
so the hardest thing yet is to call each by name.

And their warbling songs are no treat to my ears;
the sneaks hide behind leaves then taunt me with jeers.

I'll look in the tree tops, that's the best place to check;
that'll prove that all warblers are a pain in the neck!

A TALE OF TWO SPARROWS

Eighteen fifty was the year a shipment sent from Liverpool
arrived at Brooklyn Institute, a New York center with a school.

Not music, art, nor lit'rature, what did the parcel feature?
A label read, LIVE ANIMALS, so package held a creature.

Packed by hand most gingerly, the birds arrived alive,
the eight pair never really "caught," in spring they "failed to thrive."

A bigger shipment ended well, another, even better.
Philly wanted thousand birds, it said so in a letter.

House sparrows lived in dozen states by eighteen hundred seventy,
when building houses for this bird became the fad of plenty.

Many argued, Get them out! They eat not insects as we thought!
There's no more room for native birds or others that we've brought!

Digesting really starchy grains, adapting well to urban land,
they filled a new and growing niche, their numbers rose, got out of hand.

As habitats too urban turned, their numbers were declining.
In Delhi, people build for them grass houses with a lining.

∞   ∞   ∞   ∞   ∞   ∞   ∞   ∞   ∞

In eighteen hundred seventy, from sparrows of the Olden World,
were twenty or two dozen sent with feathers still unfurled.

Called German or Eurasian Tree, with insects on their dinner plate,
the birds preferred farms, parks, and towns, aggressiveness was not as a trait.

It's on one hundred fifty years, they've had no smoother sledding,
Saint Louis never far away, its range has not been spreading.

But back in Europe, some decline, while Asian population fine.

THE TOWHEES OF TURKEY HOLLOW

I met them all one day in May,
on hikes we'd call on each.

From trees in grasslands they would sing,
their songs like human speech.

The first we'd hear was "Normal Ned,"
with "drink your tea" from tree.

Next visit was to Simple Steve,
who only said, "drink tea."

Two pathways led to "Double Don,"
who sang in double notes.

"Drink-drink tea-tea" is what he said
in almost double quotes.

With "Backward Bernie," words got turned,
he always made us think,

if his songs were admonishment,
how do we "your tea drink?"

WOODPECKERS

Unique among the Aves Class are birds that peck on trees
to find the insects hidden there, uncover them with ease.

With two toes forward, two toes back, they show accepted stance
when posing for the picture books, but need a second glance.
They cling to trees with two toes up, but aiming down, just one;
the fourth toe braces off to side, where tight embrace is won.

The bird, so propped, can drill for food by using bill to peck
with quick, repeated movements at near twenty times a sec.
This rapid-fire pounding's more than any bird alive,
they make ten thousand pecks a day in order to survive.

The wood is not what 'peckers seek, they're feeling for some bugs
to catch with lengthy barb-ed tongues by many little tugs.
Attached to base of nostril right, tongue loops inside the head
and exits in between the bills to keep the 'pecker fed.

Their tongue may act as cushioning so stress of impact low.
Air pockets that are found in skull let not vibrations grow.
Between the eyes, woodpecker's skull and hyoid bone, as well,
are spongy, so absorb most shocks and verberations quell.

Near all of impact energy through body dissipates.
The bit that's left will stay in head where heat it generates.
The bill, as well, does redirect the energy of pecks.
Third eyelids at last-second blink, protect the eyes like specs.

The birds that peck know how to act, no need for any thought.
Our engineers could learn from them, if Nature's ways are taught.

WOODPECKER SPECIES

Red-headed Woodpeckers

The young wear plumage of adult, but color broken gray.
Storing nut and insect finds, it covers them each day.

Red-bellied Woodpecker

Two times I've seen the blood-red splotch that's always facing tree.
Their odd behavior might be play, but does it give them glee?

Yellow-bellied Sapsucker

Return in Spring, drill holes in line to make a well of sap
for yellow bellies, first to sip, then hummingbirds to lap.

American Three-toed Woodpecker and Black-backed Woodpecker

These species have three toes per foot, two forward, but none back;
their stiff woodpecker tails that prop make up for toes they lack.

Downy Woodpecker

Nuthatches and the chickadees make winter flocks seem odd.
Not just on trees, a downy pecks on galls of goldenrod.

Hairy Woodpecker

With larger size and longer bill, a hairy needs not wait;
at feeder, downy, hairy meet, the hairy first to sate.

Northern Flicker

It doesn't slam its head on tree, it pokes its bill in ground
and searches for the insect life that's living all around.

Pileated Woodpecker

These crow-sized birds dig feeding holes some inches deep and wide,
a few so very long and bare that insects cannot hide.

CUCKOO

I'd never seen a cuckoo,
the European kind,
but led a German walker
and had to change his mind.
The bird walk went to boat launch,
I saw a cuckoo's nest.
The walker said, "You're crazy,
unless, of course, you jest."
I, "Cuckoos in *our* country,
black-billed and yellow, too,
will build nests that look shabby,
which *your* birds never do.
"*Our* species like *your* cuckoo,
that doesn't build a nest,
is named 'brown-headed cowbird,'
with nestlings never blest."

CUCKOO TOO

Hitched a ride to Crni Lug.
Too dark to street signs see.
Wandered 'round, afraid to knock.
Where could my friend's house be?
Slept behind the church that night
on stack of roofing tiles.
Tried to find some comfort there
while gathering my wiles.
The night was cool and quiet.
Lodged there until the dew.
Crickets kept me stark awake
till called a bird "Cuckoo."

BIRD SONG

In order to communicate, birds utter different calls,
a lexicon of notes in life as told through phonic scrawls.

Their sweeter songs have other goals; each male sings in his realm
to shoo the other males away, woo females from the helm.

Her response to avian ads will carry on her race,
yet in her choice, she may begin a dialect of place.

The lyrics crooned by vying males are sung distinctively;
while judging features of each bird, she rates him on his plea.

The longest and most intricate from tiny winter wren,
just a hiccup from the Henslow's as he wobbles to his den.

Most curious from bobolink – the oingo boingo bird –
yet even more unusual when two at once are heard.

A thrush sings with both syrinx sides, a different pitch in each,
making flute-like and ethereal the lonesome male's beseech.

These mimic thrushes copy snips from songs of other birds:
grey catbirds once, brown thrashers twice, and mockingbirds do thirds.

More than a song from sandhill cranes, their courtship is a dance:
males rub on mud then bugle mates to intimately prance.

ROW YOUR CROW

Row, row, row your crow,
    murder in the air,
       big and black and menacing,
         and giving birds a scare.

           Honk, honk, honk your goose,
            skeins in V's so high,
              honking in migration to
                and from the southern sky.

              Dip, dip, dip your flight,
            goldfinch undulate,
         "potato chip," is what you say,
       black cap upon your pate.

       Soar, soar, soar up high,
     wings in shallow "V",
   scanning with your turkey head,
dead carcasses to see.

HAWK-WATCHING

Whenever I gawk,
I see a big hawk

that's almost too high
to see with my eye.

I wish it would perch
right here in this birch

and tell me its name
to settle this game.

NEIGHBORHOOD BIRDS

My sister lived in Florida, in southern part of state,
directly 'cross the tollway from community with gate.

We drove around to see inside what was so very good;
nice cars and houses, that was all, in fancy neighborhood.

Back at my sister's place, perhaps, were nature sights for me,
like lizards running hither, yon, or birds I'd rarely see.

Then, happy day, black vulture high, and hawk upon her house,
the once-endangered peregrine was diving on a mouse.

THE SIGHTING

An owl appeared and glided to a river bottom tree
and landed on the outer edge, a place that I could see.

It held its branch and looked for food upon the forest floor.
My view so good, the sight so rare, I watched a little more.

There were two tufts atop its head, but not where ears did grow;
they're buried deep in facial discs, one higher up, one low.

Such ears allow an owl at night to find whate'er it hears,
including noises under snow, and by those sounds it steers.

As well as sound, the facial discs can gather in scant light,
perceived by oddly-shaped large eyes as something almost bright.

Its eyes so big, they cannot move, an owl must turn its head,
from part-way rear and side to side, three-fourths around the spread.

With special fringing on its wings, an owl flies silently,
which helps it sneak up on its prey and kill it instantly.

Its perfect camouflage can help an owl avoid a stare.
The owl appeared and glided in, but then it wasn't there.

HIGH WAY C

Owl

One day in sunny morning-time, I walked on highway near,
to get my daily exercise, see maybe birds I'd hear.

Identifying singing birds was one of my great joys,
but on that day, so many birds, their songs were background noise.

Then, suddenly, though in the day, a barred owl made a hoot.
As dominoes, birds dropped their songs till every one was mute.

The silence broken, thereupon, each bird gave sharpest call,
communicating hawk or owl, to warn birds one and all.

Wild Turkeys

Another walk on Highway C
produced a splendid sight to see.

Impossible ten years before,
a view of turkeys, more and more.

Reintroduced Missouri stocks,
split up to form Wisconsin flocks.

The birds became good residents,
with hunts where population dense.

That day the turkeys, one by one,
flew 'cross the road till thirty done.

LEARNING TO WALK

It looked like any other bird of Great Blue Heron type,
except for ridge upon its neck that looked like extra stripe.

This year, it seemed of adult size with inch of give or take.
In height, it looked some four-foot one, the tallest on the lake.

But last year, lacked maturity, at least compared to rest.
It stomped and splashed where e'er it went since first it left the nest.

Abundance of the Glenn Lake food allowed it to survive,
despite its youthful awkwardness, it learned to stay alive.

And this year I was glad to see its graceful adult stride,
so slow and smooth, no ripples made, its prey would nowhere hide.

Mammals

Mammals

CAT FOOD

Domestic cats' carnivory
leaps from their distant past.

It's obligate, so must go on
at each feline repast.

With gut adapted, eating prey
must meet nutrition crude.

Thus bound to diet finicky,
a house cat seems less rude.

If captured outside in the wild
or served up in a dish,

or meal of mouse that's caught by cat
or can of tuna fish,

the feline diet oft can cause
a curious reaction

that coincides by senses keen with
turned-up-nose olfaction.

While kittens play with other cats,
adult cats hunt alone,

they use their pet-hood social skills
to independence hone.

A DOG'S LIFE

As if in pack with neighbors' dogs, they run, run, run;

they love to play with humans in the sun, sun, sun;

the game of fetch with Frisbee is so fun, fun, fun;

a dog might chase its tail around till spun, spun, spun;

a hunting dog still charms when there's no gun, gun, gun;

dogs even eat the wiener and the bun, bun, bun;

the sled dogs in a team pull near a ton, ton, ton;

dogs keep intruders out, allowing none, none, none;

by loyalty the owner's heart is won, won, won;

a sleeping dog is truthful, not the pun, pun, pun.

RARE SIGHT

I headed north with travel friend, near Forest Nicolet.
We went to see the meteors, so drove up one clear day.

While up there we decided we'd each ride a rapid raft,
so made our way to river wild with rubber watercraft.

First drop of river threw me out, again and yet again.
Fourth time the charm, no body harm, put rafting in my ken.

But next came shallow river bed, where boulders stopped the boat
so that I had to leave the raft to help the vessel float.

Then roaring rapids, quickly passed, forced many daring swerves,
awareness peaked, my muscles taut, with frayed and tingling nerves.

The river flattened, water calmed, we moved into a cove.
With gentle waves along the shore, we saw a tiny grove.

Emerging from the little copse, it stood in fullest view
and lingered for a moment's time; my eyes saw something new,

like fifty-some around the state, worn collar radio.
Not husky dog, instead we saw endangered lupine show.

QUARTER AFTER DARK

The clock read quarter after dark,
guests started to arrive.
Some unrelated flying squirrels
will den together, thrive.

The Tension Zone from plant range maps,
can show where dwells each kind,
the larger northern flying squirrel
or smaller southern find.

They may be seen in skies made dim
in glow of distant light.
They forage acrobatically,
by sniffing or night sight.

Despite their name, they do not fly,
these mammals modified.
Time sewed for each patagia,
courage the nerve to leap and glide.

Patagia (*puh-TAY-jee-ah*)
        Plural of patagium.
        The fold or flap of skin membrane
        that connects the forelimbs and
        the hind limbs.

Tension Zone
        A region dividing the northern primarily
        coniferous forests from the southern and eastern
        broadleaf deciduous forests in Wisconsin
        & other Midwestern states.

Note: As of 2024 the Southern may have mostly pushed the
        Northern out of the state, due to climate change.

BATS

Like Nash, I rather like the bat,
its high-pitched chirps are never flat.

Too high to hear for humankind,
all bats, not blind, by echoes find

the flying insects that they eat
with wayward skeeters for a treat.

Erratic flight to catch the bugs
that veer too near their mouse-like mugs.

With furry bodies, wings of skin,
stretched tight between long fingers thin,

some types of bats migrate in waves,
while most bats hibernate in caves.

The vampire bat is never found
where northern species hang around,

but places south, like Mexico,
and movie screens with vampire show.

GROUNDHOG DAY

Three hundred sixty days a year, plus four and sometimes five,
woodchucks are not that often seen and rarely seen alive.

The few who see 'chucks in the wild consider them a pest,
except on Feb the second, when they run their weather test.

'Chucks claim the name of "groundhog" then, which grants them god-like traits:
prognosticating climate and predicting vernal dates.

By legend, groundhogs leave their dens, each checking if it's made
a terrorizing silhouette or other scary shade.

If frightened, they return to hide from Winter six weeks more;
if Sun's not out, there's early Spring in shy of six weeks four.

On Candlemas the groundhogs played this shadow-boxing game
till one, because of local press, achieved the greatest fame.

Punxsutawney Phil is roused from hibernating slumber,
and asked, while still half-frozen, for the weeks of winter's number.

Most groundhogs end their torpor one month later in the wild,
but winter-weary folks can't wait for weather that is mild.

PORCUPINE

Each year we count our breeding birds
to note the rise or fall

of numbers in their habitats
and help when they get small.

To Hiawatha first we went,
where counters were before,

to chart how data changed from past,
the science not a chore.

Arising much more difficult
at quarter after four.

Then we saw life crepuscular,
a porcupine afore.

I rushed to pet the animal,
just missed the needled beast.

I tried to show bravado,
but pet attempt was least.

OTTERS

Our survey done, the count complete,
we wanted next a little treat.
More wildlife seen in refuge near
might fill our eyes and give us cheer.
We therefore changed our pathway back
and took the Seney Refuge track.
Indeed, we saw in open pond
what made us smile with feelings fond:
a raft of otters set to play
by treading water in the bay.
They looked like people in a pool.
Do people look like otters? Cool .

Other Animals

Other Animals

MUSSELS

Without an eye, they watch Life's flow, the Mississippi clams.
At first, unbounded, then impounded by many locks and dams.

The yearly rhythm, flood and drought slows not the growth of shells,
but darkened rings denote the time of winter weather spells.

Incurrent siphons bring in flow for mussels' life-long sip
to filter plankton, history from Mighty Mississip.

Beneath the mantle found within each shell that's made of lime,
is nacre lustrous, rich and smooth, sought after over time.

Ere plastics fastened all our clothes, pearl buttons were the trend,
this widespread use of mussel shells had put them near their end.

Then numbers rose, with types diverse, life cycle seemed secure:
eggs fertilized, glochidia, benign on host mature,

young mussels drop from fin or gill, beginning life anew.
The habitat that mussels need supports the host fish, too.

The latest pressures mussels feel are species introduced,
disrupting native mussels' lives till numbers are reduced.

A mussel, sessile, anchors foot and moves itself a smidge.
Its body leaves a dragging trench, on either side a ridge.

Our eyes see zebra, quagga clams move at invasive speed.
Must slow them down to native pace, change something that they need.

MUSSELTOWN

The clams I've seen in Mussel Town left stories in my mind
of characters from olden times in shells that we could find.

The shells I wanted most to miss and avoid Achilles' fate
were **heelsplitters**, **creek** and **white**, though cutting power not great.

I traveled next with Gulliver and searched for tiny **Lilliput**
by land and sea and ev'ry where that I could go by foot.

Lenses from my **spectacle case** made it easier to see,
but if I used some **snuffbox** snuff, I might then over-see.

That night we lay in **mucket** bed and got so full of muck,
we used a **washboard** for our clothes, so didn't have to chuck.

I had not a **fat pocketbook**, no **floaters**, **strange** or **flat**.
Our **papershells** were **fragile**-named, but strong in spite of that.

I asked my fellow traveler for someplace we could go
where we could spend the whole long day with not a lot of dough.

The character said "zoo" for sure, then later, maybe, "feet."
And there it was, between my toes, a laminated sheet,

a guide to river mussel shells' menagerie of names,
not quite a zoo, but close enough to what the story claims.

Fawnfoot, deertoe, elktoe, too,
buckhorn, bullhead, more for you.

Elephant ear with **monkey face**,
a **pigtoe** seems not out of place.

BACKWATER BATTLE

I saw the struggle, life and death,
a savagery that took my breath.

One fiercely fought, its life to save,
on every gently flowing wave.

It thrashed and took an upstream tack,
the fish snake-bitten on its back.

The snake and fish, with raging might,
swam in their fury out of sight.

The rivals soon in weakened state
succumbed to river current fate.

Descending past, the bitter foes
dragged out their drama toward its close.

The perch could not its burden shake
nor predatory death-grip break.

The water snake persisted more,
yet river washed them both ashore.

SALAMANDER

While camping on a trip alone in tranquil part of state,
I pitched my tent near gentle stream, the fire would have to wait.

Flowing water soothed my ears, so full from noise that day.
Some crickets chirped, grasshoppers buzzed, but water found sleep's way.

Then Dawn began to inundate the darkness of the sky.
I grabbed the flap but slowed my pace, no reason yet to hie.

Unzipping slowly as I could, revealed a hidden sight,
a spotted salamander near, its visit a delight.

It stood stock still, quiescently, not hidden by the dark.
So common yet so rarely seen, this denizen of park.

It left no sign of movement there; no tiny mark on trail.
A short time on, it had not stepped, nor even flicked its tail.

I watched much longer, mesmerized, it did not stir or flee.
Until, at last, I made a noise, it left and set us free.

The herp, from its biology of motionless defense;
me, from waning wonderment that still held me in suspense.

FROG CALLS

In springtime, frogs begin to call, in ponds and wetter lands.
Before each call, their lungs inhale, their vocal sac expands.

This air is forced from sac to lungs, but never is it breathed.
The frog or toad tight lipped remains, its tongue is ever sheathed.

When air flows o'er their vocal chords, anurans sing the songs
comprising aural signatures, distinct from that of throngs.

The calls establish breeding grounds, the songs attract new mates;
frogs in a chorus pattern sing to see how each male rates.

The competition orderly, one frog and then the next;
each solo artist sings his song, described in coming text.

A Wood Frog makes a croak from home
while Chorus Frogs strum on a comb.
The Peepers' peeps are next to spring,
and sound like distant sleighbells' ring.
A Leopard Frog rubs tight balloon,
a Pickerel then snores in tune.
For many seconds trills a Toad,
before Gray Treefrog's razzing flowed.
Then Cope's Gray Treefrog buzzes back,
as Cricket Frogs two marbles clack.
The Mink Frog trots on cobblestone
and Green Frog twangs loose banjo tone.
The last anuran call from lake
is foghorn sound that Bullfrogs make.

A SURPRISE

Toward colors of the forb I crept, to photograph the bloom
with flowers' petals woven low in Mother Nature's loom.

For land ne'er touched by human hands, nor sprinkler, nor mower,
I wanted level nature shot, so knelt down even lower.

But when too close to Earth I leaned, a milk snake buzzed its tail
in driest leaves to scare me off. In that it did not fail.

I knew it was not rattlesnake, but that was just my smarts.
I yelped and fled from unseen snake as told by heart of hearts.

ANOTHER SURPRISE

Each week I 'd lead a hike or two or three
at camp for those with disability.

The Badger Campers followed me on trail,
we slowed the pace if someone was not hale.

I stopped the hike when something made a sound.
Could not mistake this rattle from the ground.

"We must go back. A big tree blocks our way,"
a counsellor, quick-witted, thought to say.

This cleared the path before the snake could strike.
Without that snake, we finished off our hike.

HOGNOSE SNAKE

In early times, amphibians developed noxious skin;
for generations, toxic toads had evolution win.

By force of Life, not conscious thought, snakes started to withstand
the bufotoxic chemicals secreted from a gland.

But toads have yet another ploy to keep from being prey
of hognose snakes that hunger feel and catch a meal to slay.

Toads bloat themselves too big to eat, yet fight not always won;
the serpent has rear fangs to use to pierce, deflate – toad's done.

When hognose snakes the prey become, they differently behave;
their acts confuse their predators and scare away the brave.

This snake may flatten out its neck and feign a cobra hood;
behaviors shown by hognose snakes are neither bad nor good.

Their instinct makes them puff and hiss, the moves an adder makes.
Without intent, the serpent sounds like other venom snakes.

The final deed this snake may do is something 'possums try:
if faced with certainty of death, a hognose snake will 'die.'

With belly up, this snake will rest, might even droop its head:
but if it moves, will freeze again to show that it's still 'dead.'

TURTLE SHELLS

A turtle with no shell attached would not a turtle be.
Its shell encumbers not its life, but neither sets it free.
Protective plates upon its back, reduced mobility,
adaptation set the course for shell anatomy.

Its scutes as plates of armor serve to shield and thus defend
both body and the skin beneath, on which the scutes depend.

The upper shell, or carapace, makes turtles look distinct
while plastron, or the lower shell, by flesh and bone is linked.

Box turtle species have a "hinge" across the bottom shell
affording them a little "box" in which they fit quite well.

By pulling in all legs and head, they hide from predators;
then pressing plastron, carapace, will open up its "doors."

The Blanding's turtle (semi-box) has "hinge" that's used in fear
to cover head and legs in front, but not the ones in rear.

An adaptation found in some freshwater-dwelling kinds
is soft shell with consistency of firmest citrus rinds.

The Swinhoe's softshell, rare on the Earth, can live a hundred years;
so nearly all were hunted out, now real extinction fears.
A male lives in captivity, a female, Viet Nam.
These two must be together bred so she can be a mom.

LAYING EGGS

"Our" turtles laid their eggs each June,
in lawn of where I worked.
I missed their laying many times
while they in sunshine lurked.

They were not ours, but used our land
to dig their holes for eggs.
As other turtles did the same,
one signaled with her legs.

Demonstrating Nature's power
when at a little road,
she stepped her foot and stopped a truck
while other traffic slowed.

Not caught or killed, a turtle stays,
and lives from day to day.
Most often, Mother Nature wins,
when given right of way.

Insects

Insects

HONEY

I pet a bee whene'er I dare, I try not to recoil,
for in this clumsy way I try to thank her for her toil.

I watch the workers, one by one, across the meadow fly;
the forage bees have said with dance that there the flowers lie.

Some flowers' petals guide the bees and show them where to light
to gather nectar from each bloom, and pollen, dry and bright.

With baskets full of pollen food, they make a bee-line back,
the nectar sipped by worker bees still held in special sac.

By work of mouth an enzyme turns the nectar each bee brings,
to honey thin that next is fanned as workers flap their wings.

The thickened sweet is stored in combs of hexagon-al cells
and capped with wax for winter food, creating honey wells.

The royal jelly's made and fed three days to all young bees,
but on and on to larvae crowned "queen mother," if you please.

By mating with the male drones, establishing her brood,
she gives them life, then I can share their produce and their food.

SPIDER WEBS

All spider types make protein silk, as liquid, pure and clear,
then load it in their spinnerets, the structures on their rear.

The polymer extruded firms as strong and tough as nails,
yet flexible, with tensile strength, the gossamer-made rails.

The architect first takes a chance, throws caution to the wind,
letting go a test balloon with end of tether pinned.

The silk balloon, made sticky first, floats off on gentle breeze;
balloon adheres to leaf or twig, is tightened then with ease.

Back and forth across the bridge to strengthen and inspect.
From anchor points, the frame is built in plane that stands erect.

First, central hub then radii are strung with silky strands.
Orb weavers next the spirals make, with tiny hook-like hands:

the dry lines first, from inside out, so spiders don't get caught,
then "capture spirals," outside in, with spider glue are fraught.

Orb weaver species make their webs in every shape and size.
Their beauty makes us marvel at the engineering prize.

So lovely and so intricate, it's hard for human thought
to fathom spiders' silk reuse; they eat what they have wrought.

BUTTERFLY PARK

My sister took me to a zoo of speciality.
No mammals, birds, or fishes there for visitors to see,

just skippers, moths, and butterflies, the lepidoptera,
from tropics and the temp'rate lands near southern Florida.

I walked into enclosure huge, took third step then aback.
"Was that a reddish arc I saw on forewings brown and black?"

With many thousand butterflies from Tropics in the room,
I saw Red Admiral Butterfly on tiny Nettle bloom.

My seeing butterfly I knew, spurred me to look for more,
for reading all the species signs seemed like a mammoth chore.

By looking for the common ones, I thought I'd better fare;
with only time for quickest glance, I'd have no time to stare.

A Comma or a Question Mark? I wasn't really sure.
Those butterflies look much alike, distinctions sometimes blur.

The Swallowtails were largest there; Blue butterflies were least.
Their rarest cousin was not there; its habitat decreased.

My eyes grew larger, most impressed by being in a zoo
where butterflies exalted were and common given due.

MONARCHS

From eggs laid under milkweed leaves, the caterpillars grow,
devouring the bitter leaves by munching off each row.

The bitter taste is glycoside, a poison to the heart,
tolerated by the monarchs, though toxic from the start.

The only job for larval form is eat and eat and eat.
For dozen days they do this well and never skip a beat.

Fifth instar of this larval stage will change to butterfly
once it's emerged from chrysalis and after wings are dry.

The western monarchs hibernate, the eastern fly in fall.
They overwinter Mexico, three thousand miles in all.

Five generations fly return, they break it up in parts
and fly a route that's new to them without a map or charts.

'Our' monarchs use two habitats, two places where they dine.
The loss of milkweeds here *and* there has led to steep decline.

With climate change, predaceous birds, and maybe hit by cars,
the monarchs are not kings and queens, but butterflying stars.

CRICKETS

When male crickets want to mate,
they rub their wings and stridulate
by scraping 'scraper' 'cross their 'file'
a nearby female to beguile
with buzzy chirps in charming song,
again, again, the whole night long.
Most people find that crickets soothe
but others wish harsh chirping smooth.
Cave crickets, "mute," of wings devoid,
make some with basements feel annoyed.
Crickets chirp from planting rows
till harvest days come to a close.
These bookends make the cricket sage
to those who shrine them in a cage.
At times in Western countries, jeered,
throughout the East, they are revered
for all the chirping songs they sang
from time of Dynasty of Tang.
Where crickets dwelt, good fortune found,
but here all live upon the ground.
Inside, a cricket stops its chirps
to signify invading perps.
Outside, crickets get a feel
for temp'ratures they then reveal.

FIREFLIES

When fireflies feel mating urge, males head to humid field
to cruise low vegetation haunts and signal lanterns wield.

Within these lamps, the chemicals react in mystery;
luciferin, luciferase make eerie light we see.

It's yellow, green, light red, or orange, one species even blue.
These magic 'flies withhold their light till dimness is their cue.

Then oxygen in tracheoles is fuel for bright, cool light;
without O_2, reaction stops, and 'flies blend in with night.

The bits of light are flashed by males, in pattern of their kind;
in Smoky Mountains, unison, in Rockies, dark 'flies find.

When ladies of the lightning bugs see gents with finest glow,
they wink assent by flashing lights that match the picked male's show.

The femme fatale *Photuris* 'bug fools males of other kinds
by mimicking their blinking codes to eat lured 'bugs she finds.

Pairs trip the light fantastic while at Nature's 'bug-lit ball;
their messages mean naught to us, yet still they do enthrall.

LEAFCUTTER ANTS

Some thousands of chambers for queen and her kin,
but only her daughters will live there within.

From dozens to millions, a colony holds
the ants for all jobs in production of molds:

cut leaves while they're green, store in beds underground,
make mulch for the fungus to which they are bound.

The labor's divided by size for each caste,
the soldiers are largest, while farmers are last.

Crop-growers, as smallest, take care of the young;
they vaccinate larvae and take out the dung.

The larger ants sometimes will help clear a trail,
if an object too big, and smaller ants fail.

With leaf-cutting roles so ingrained in the ants,
they're essentially movers of rubber tree plants.

With branches as pathways, the better she copes,
with possible tasks, to fulfill her leaf hopes.

Note: Poem alludes to song *High Hopes*

MAY FLIES

Other species have a "hatch," but none is quite so grand
as that made by Ephemera, the largest in the land.

So big in fact, emergence seen by RADAR, set for rain,
which estimates swarm numbers to show if they grow or wane.

Four 'flies crashed into flailing arms; one hit me in the ear;
a few alighted on each leg; a dozen flew too near.

The living cloud appears at dusk and congregates near lights,
lays waste upon the river towns and stays till dark of night.

A swarm begins near water source that's clean enough to "hatch,"
a thousand million mayfly duns, enormous breeding batch.

The male lives but a day or two then goes on land to die;
the female lives a shorter span; she mates while in the sky.

With mouth parts rudimentary, adults don't bite or eat;
the larval nymphs eat food enough, life cycle is compleat.

Drab duns become more colorful as they to spinners change,
this second flying insect stage, unique and rather strange.

With no reserves, she lays her eggs from surface, on her back,
they gently sink, adhere to logs, or fall to bottom black.

The eggs transform to naiads (nymphs) one time or manyfold,
the nymphs live then at least a year; this life is rarely told.

While laying eggs, the mayflies are like "sitting ducks" for fish,
who easily can eat their fill of tasty insect dish.

Fly fishers tie their flies to look like mayflies to a trout,
then act as insects in a swarm to catch a fish to tout.

TWO TALES OF A GRASSHOPPER

Grasshoppers of the short-horned kind
near Earth on grassy plants are found,
to eat the food, politely dined,
but not yet any dregs around.

If arid lands the rains do sate
they grow green plants most everywhere
then feasting comes and seals their fate,
till not a plant is anywhere.

Some 'hoppers into locusts turned
by serotonin held within
that powers frenzy till adjourned
to make a raucous chewing din.

But, worst of all, when locusts touch,
which, in a swarm, they can't avoid.
In parts of world, they're hated much,
but in US, we're just annoyed.

WATER STRIDER

Jacanas walk on lily-pads
and houseflies walk on walls,

squirrels can climb electric poles
and cave fish, waterfalls.

How sacred the relationship
that forms a living bond,

imbuing water-striders means
to walk across the pond.

The surface tension water has,
supports the insect's 'feet,'

the strider so adapted learns
from many times repeat.

With wings, the strider's unconfined,
yet stays in habitat

to take advantage of its traits
that work where water's flat.

Plants

Plants

LICHENS

On the Isle Royale and to its north west,
where looking for lichens is always the best,

with so many lichens of so many kinds,
people go out on hikes and count up their finds.

They find them on trees, and bare ground, solid rock,
in morning and evening, midday on the clock.

The flat, crustose growth form that some lichens take
produces same surface old spray paint may make.

The foliose lichens appear to have leaves,
while fruticose lichens reach up with long 'sleeves.'

The shape of a lichen comes from fungal part,
with acid it uses to break rocks apart.

This rock-breaking role of the lichens is key
to add to nutrition to soil that is free.

The fungus is shelter for algae to live
and make all the food that they're ready to give.

This housing-for-food keeps communities strong
in complex arrangement where life forms belong.

Next time you're in Nature and find yourself hikin',
remember to stop and reflect on each lichen.

FERNS

I took a stroll on shady bluff between two types of ferns,
while one could follow any path, one only downward turns.

Each species reproduced itself by spores and not by seeds,
so cloaked the cliff with lacy green, not flower-colored weeds.

Though tiny, spores hold DNA instructions for the plant.
Spores moist and sheltered, germinate, but otherwise they can't.

Two generations alternate, the "fern", or sporophyte,
and overlooked gametophyte, a fingernail in height.

The ferns along the trail I trod regenerated there,
without the spores of other ferns nor reproductive pair.

Ferns do it vegetatively, each bulb is then a clone,
a tiny bulb along plant's keel, by Mother Nature strown.

Dried bulblets move by gravity, along the bluff-side edge
until they meet an obstacle, stopped by a little ledge.

The other fern can 'walk' the Earth, re-rooting forms its 'grip',
then takes a step, a growing stride, re-roots again at tip.

EVERGREENS

In Winter, when the broad-leaved trees
are dried by Nature's frozen breeze,
leaves fall to Earth, where they will die,
but not from cold, just much too dry.
With waxy coat and needle shape
and stomates rarely held agape,
a pine survives the winter bleak
with moisture never more than meek.
All pines, like broad-leaved trees, will lose
the leaves that act on Autumn's cues.
Half the needles fall each year
from white pines, leaving browse for deer.
By species certain portions fall:
a third, a fourth, a fifth, or all.
The last is larch or tamarack,
its late-year needles pigments lack.
The first four show, with winter green,
the life in death by Faith that's seen.

OAKS and MAPLES

A maple's almost only tree that in its shade can grow,
while oaks in shade of any tree have growth that's rather slow.

So maple woods are deep and dim with mostly maple trees,
yet here and there a basswood tree gives pollen used by bees.

The oak woods bright and open are, competitors abound.
Black oaks prefer the drier sites, few species there are found.

When maples lose their leaves in Fall, their foliage is bright.
That matters not in Nature's eyes, She wants for food to bite.

October maple leaves aground are good, so rarely seen
for Decomposers on the spot can pick the leaf stems clean.

But oaks hold tight their leaves and get the last nutrition bits,
until November, some trees, March, before their leaf-fall quits.

The maples more nutrition lose; it's in the leaves they drop.
Since maples grow in their own shade, those trees come out on top.

The oaks are much more frugal of nutrition in the soil,
October leaves less nutritive, oaks are the maples' foil.

FRUITS

It cannot be, you must be wrong!
I'll show you books. It won't take long.

It sounds not right. How can it be?
Come take a look. Let's have a see.

That's not the story I've been told.
As scientists, we must be bold.

The question is no longer moot?
Tomatoes are a type of fruit!

Is it the only food like that?
When fruit puts on a veggie's hat?

And takes the culinary name?
Even when it's not the same?

So many fruits from ovaries,
like peppers, olives, okra, peas,

still considered vegetative,
even though parts generative.

What rhubarb parts meet baking goals?
The stalks of leaves, or petioles.

Whatever fruity taste belief?
Not fruit, but stems of each large leaf.

Replaces fruit by those who bake,
and pinch off buds to crumbles make.

This sour produce just may be
what goes down hard **for you and me.**

GARDENING

Gardening's Goals
reap, eat,
repeat

Gardening's Gifts
peace of work
from
piece of work

Stories

A MOMENT WITH GRANDMA

While I was young, came Grandma to our door.
She'd visit us a week, or even more.

With Dad, she spoke Croatian for her lore,
we nodded at her many days of yore.

And then one day outside was something new.
She ope'd the curtains for a better view.

There framed a picture of uncounted birds,
narrated by her best Croatian words.

From one horizon to the other flew,
in speckled sky that otherwise was blue,

enormous flock of birds, all colored black,
a million minds all on a single track,

wherein no thoughts of territories grew,
migration, only, as the blackbirds flew.

For next half hour, Grandma bit her tongue,
and in that moment, all who watched were young.

MR. SALTZMANN'S GARDEN

On sunny days, we used to walk the hundred feet or so
to Mr. Saltzmann's garden shack, Mom rarely told us, "No."
We never went inside his shack, he said not, "Yes, you may."
His hand pump, though, was free to use, we drank from it each day.
The garden, one full acre large, was where he grew his crop,
raspberry plants in such long rows, they never seemed to stop.
He labored hard, but took a break whenever we appeared.
We never ate his berries, though, the produce that he reared.
Just four was I, my sister, five, when first we went next door,
where all we ever did was talk and started not a chore.
I've pondered what we talked about, but don't remember what
the details, even subjects were, by Mr. Saltzmann's hut.
He planted in the spring each year, so reaped on summer days
and weeded when he got the chance, with rarely time to laze.

One year in latter summertime, my father said one day,
"Let's go see Mr. Saltzmann now, he lives not far away."
Dad buttoned up a button-down, Mom donned her favorite blouse,
five minutes' drive, Dad parked the car at Mr. Saltzmann's house.
He greeted us outside his home and introduced his wife.
We looked inside their living space and saw his other life.
She wore a simple dress she sewed and he a bolo tie.
With all the fancy things about, I suddenly was shy:
I didn't want to break their things, so hid behind Mom's skirt
and didn't touch the crystalware, I brushed, instead, my shirt,
the pieces of fine furniture, alone or matched in sets,
the tables, lamps, and carpeting, the fancy cabinets.
One cabinet held curios, which fascinated me.
Although we looked until we left, I wanted more to see.

He never showed his house again, nor sat outside his shack,
for he was in his 80s then, soon after, life went black.
His words are long-forgotten now, his actions are my guide:
When interrupted, take a break, hard work will give you pride.

156

FEEDING BISON

The herd was tens of millions strong when money changed no hands,
yet currency of bison skins brought more and more demands.
Once ev'rywhere but near the coasts, no thought they'd disappear.
Then slaughter inconceivable, two million in a year.
Two dozen left in Yellowstone created public fears.
No longer being targeted helped dry the public's tears.
Restrictive hunting laws enforced, the bison breed like deer.
Now herd is half-a-million head, their comeback makes us cheer.

We drove all day on family trip in station wagon car.
No phones or tunes or videos, the drive seemed very far.
We missed by minutes open hours to see the bison park.
With silver tongue, Dad made a deal that wouldn't miss the mark.
We'd ride along with staff and feed the bison herd a snack
with Mom in middle, Dad on right, and children in the back.
A worker drove us in his truck to middle of the herd.
"Don't touch the horns!" he sternly warned two times and then a third.
Its head as big as all of me, I feared for being bit,
but there, inside the pickup truck, I didn't want to quit.
Such herbivores ate only grass, so it would do no harm.
Yet when I fed it pellet food, it sucked up half my arm.

*This poem is from a different era.
It is not wise, nor encouraged to get close to bison.
Do not try to feed them, nor touch them.*
(Luckily no harm occurred.)

NETTLES

One day while at the garden bed, when I was just a child,
my father tried to teach me weeds encroaching from the wild.

"Come over, Son, and feel this plant, with nettle for a name."
My tender fingers burned and itched, as if too near a flame.

Much later, while a nature guide, I studied nettle plants,
which have the formic acid found in painful biting ants.

The hollow hairs of stem and leaf break off and then inject
their acid into skin laid bare, where clothes do not protect.

Laportea, wood-nettle, grows in shaded, wooded stands,
but stinging nettles, *Urtica*, in open, vacant lands.

With cooking, nettles lose their sting, are safe and edible.
Their simmered leaves and cooking broth each taste incredible.

While leading hikes, I always warned to touch no nettles raw,
nor eat them without cooking first, lest feel a throbbing maw.

One day while hiking on his trails, I told this to my dad,
who bested me a second time, ate nettles and was glad.

GOOSEBERRIES

One day while swinging happily, when I was just a lad,
I heard him call from 'cross the lawn, so ran to be with Dad.

"They're ready, Son," he smiled at me. I peered around his knee
then stared in horror at the shrub, "There's prickers!" was my plea.

Dad said the berries tasted good and gave me some to try.
I paused and backed a step or two and then began to cry.

I whined and whimpered endlessly; I did not want to eat
the pea-sized fruits with pointy parts, though Dad called them a treat.

He said again they would not hurt; my heart and mind were torn.
He ate a few to prove them safe, I touched the longest thorn.

"The fresh new spines are soft, not sharp," said Dad's assuring voice.
I sniffled once then wiped my tears, looked up and made my choice.

The unripe fruits had tasty crunch, two acids made them tart.
Dad always knew I'd say someday, "I loved them from the start."

TELEVISION

When Dad got home to dimming sky, he changed from clothes he'd worn.
We watched our shows, my sis and I, cartoons were in the morn.

Dad worked, with never being seen, he snuck behind the house.
While glued to television screen, we were quiet as a mouse.

Dad stole back in and hung his hat, then kissed us on our heads.
When shows were done, Mom gave a pat, we climbed into our beds.

Next morn, Mom pried us from the set, she heard me rant and rave.
Outside stood Dad, his coat still wet, in drift he'd dug a cave.

The snow of cave trapped bits of air, so transferred little heat
to insulate our little lair and make it be a treat.

We played all day, were never cold, protected from the wind.
Our TV plans we put on hold, my father only grinned.

SKIPPING STONES

I learned to skip stones flat and smooth
from older, smarter cousin,

but even with his tutelage,
I couldn't bounce a dozen.

My cousin, Don, skipped other things,
whatever shape they'd take.

A marble once he even threw,
and made it skim the lake.

I tried not many tries and thought,
"With round I'll not excel."

I tried again with flatter stones,
but didn't do as well.

All week we pitched stones in the drink,
his numbers out of reach,

so next time we were at the lake,
we simply walked the beach.

BULLHEAD FISHING

I was never fond of fishing, not walleye, perch, or bass,
or any fish caught from a boat; I'd want my feet on grass.

The only kind of fish I'd catch were sunfish in the sun,
we'd use for bait so chopped them up, before the day was done.

Emerging later, 'fore the dark, to set the lines we'd tend,
each cousin watching poles at hand until the very end.

Once poles were set, we'd go about the business of the night,
we'd check the poles for bait-less lines or bobbers out of sight.

Then, best of all, between the checks, were cards and cards and cards.
We bet no money, just had fun, no need for any guards.

The cards were conservation ploy to think about the lake,
and bottom-dwelling fish therein, precautions that we'd take.

With "whiskers" soft a bullhead feels what passes by its head,
but spines near fins are hard and sharp, may injure you instead.

Low oxygen and turbid pools are bullhead habitat,
they'd likely choose another place, but no one asked them that.

NATURAL HISTORY

Our fam'ly owned some hillside land, its acres numbered ten.
We didn't live too far away, I biked there now and then.

With grass aplenty, mostly brome, there wasn't much to see
of native plants or what's entailed in nat'ral history.

In storm one year tornadic winds took out an aged oak.
My father and his friend next morn cut all before I woke.

"Please let me count that red oak's rings before you haul away."
One hundred thirty-six I got, with checks it did not sway.

The count reflected age at death, enabled me to try
to understand what lived nearby and grew beneath the sky.

The point on Earth the tree would sprout was near a settlement,
the second in Wisconsin-land, where many Norsemen went.

The tree we'd cut, not yet a seed, would grow with oaks of white,
and in the shade of older trees, where sunshine not as bright.

In eighteen hundred thirty-nine, some took their land for free,
replacing nearby village of the Potawatomi.

In middle of the century, once cholera had struck,
'twas named the Town of Norway and ancestral name has stuck.

The tree was not susceptible to cholera or flu,
so there, above Wind Lake by name, the stalwart red oak grew

until my father owned the land where history was writ
in ev'ry leaf of ev'ry tree, so learning never quit.

For hundred plus three dozen years it held its secrets well.
I read its years at end of life, no spring-times still to tell.

THE LAND BEHIND OUR YARD

In pasture-land behind our yard grew many hickories,
and one bur oak, so big and stout, it never bent in breeze.

Each year in Fall my sis and I would gather hickory nuts.
We'd always try to skip this chore with ifs and ands and buts.

It never worked, we'd still collect, but after Labor Day.
Dad said, "Don't eat the early nuts, they're likely not okay."

A 4-H friend of mine and I once tried to capture Time.
We knew old trees grew wider still, once big enough to climb.

We had to know, how do trees grow? At rapid pace or slow?
Or do they stay at smaller size and then put on a show?

To find this out, we got a tape and measured 'round the oak,
which gave a clue so loud and clear as if it once had spoke.

The tree was thirteen feet around, and still nine inches more.
In fifty years competing there, upon the forest floor.

I saved the slip, lost track of friend till he bumped paths with me.
New measure sought, I wondered; how much bigger it might be.

Alas not so, five inches lost, bark missing now a bit;
lightning, lean years, competing growth; at least, it has not quit.

GOLDFINCHES

My father never liked to change; he gladly lived in ruts,
but knew he should not miss a chance before a doorway shuts.

A thrifty man, within his means was how he lived his life,
was always thrilled with gifts received, best of all, from wife.

She knew always what he wanted (a thought before he did):
a modest gift, complete with task for her to keep it hid.

He couldn't wait to open up the presents that she bought,
not socks, nor ties, nor underwear, nor other things he sought.

That year my mother's gift for him and thistle-eating birds,
a feeder finches would enjoy, but not our squirrel herds.

My father set it up anon, along with Nyger fill.
That day two finches stopped to eat and gave us all a thrill.

Dad pondered the supplies on hand, a twinkle in his eye;
he had enough materials, to make a test to try.

Next day he went to hardware store to buy more things he'd need
to make the feeding stations for the fourteen birds we'd feed.

POISON IVY

A mate came here from 'cross the Pond,
where grows no poison ivy frond,
which caused a hike to stretch our bond.

I showed the plant and took the lead,
but he did not my warning heed,
and brushed against the itching weed.

"If woody stems and leaflets three
on creeper, shoot, or vine you see,
the plant, and all it touched, do flee."

But next day no reaction seen
from ivy with red glossy sheen
or leaves of Summer, dull and green.

Again the next, no rash from plants.
Emboldened now, he took more chance,
the day a hole ripped in his pants.

Urushiol was wiped on thin
to newly unprotected skin
on rear of leg, above the shin.

Now twice exposed to ivy oil,
immunity began its toil
and made his fluids start to roil.

My English friend, intrepid Brit,
showed several days tremendous grit,
for all the while, he could not sit.

A TRIP CUT SHORT

Each day we pedaled south or west, depending on the wind,
until we got to foothill height, where higher air had thinned.

My friend went back. I biked alone at least a thousand miles.
The little hills outside LA were least of biking trials.

But try and try and try again, I could not ride to top.
Smog beat me down and slowed my climb and three times made me stop.

Emissions are much better now, as low as if most biked,
let's take it down another notch, as low as if all hiked.

GETTING THE JOB AT WYALYSING

For weeks, we planned our camping trip
to Wyalusing Park.
We'd ride atop our bicycles,
three days until the dark.

Three more around the park for fun
and three for riding back.
We never used the plan we made,
'twas punctured by a tack.

Dad's tires from Sears were oddly sized,
the spare just did not fit.
So Dad called Mom. They hatched a plan.
We might not have to quit!

If Mom still did not want to camp,
she'd fetch us, we'd go home.
But if she wished to join our jaunt,
we'd ride no wheels of chrome.

Once she found out she would not ride,
Mom knew just what to say.
I've written much about my dad,
but Mom got me to stay.

And stay I did, twelve seasons strong,
first month at maint'nance fee.
I led the hikes and programs, too,
first did them all for free.

FIRST BIRD WALK

I poorly planned my nature walk for first day in the park,
we watched the birds along the bluff, but hike soon left the mark.

Each step became more treacherous, no longer on the trail,
my leadership had showed us naught, but Nature did not fail.

Some dripping water showered birds, one green with yellow breast,
the other bird less camouflaged, with striking colors blest.

The red and back of male distinct, their contrast weakens knees,
but female held to different scale, her subtler colors please.

The male began the toileting, alighting rather low,
adjusting his positioning to dip beneath the flow.

Clear water splashed upon his head to make his body clean,
he perched upon a nearby twig and thereupon to preen.

The female scarlet tanager, at basin after mate,
used her time to cleanse her skin and comb her feathers straight.

Each sat before the fountain drips and took a final turn
then followed Nature, formed a bond, eloped behind a fern.

HUMMINGBIRD TRAP

The smallest bird I'd ever seen, its back and wings metallic green,

ensnared within a monstrous trap with, at one end, enormous gap,

held hostage by the thought ingrained that flying up is freedom gained.

Without delay and not a word, I raised a staff to help the bird.

On end it perched repeatedly; I dipped the pole to set it free.

Each time it got near open space, it flew to roof-board closed-in place.

The hummingbird began to fade, once buzzing past, it now was staid.

The little bird had little time; with bag in hand, I made my climb.

Ascending to within arm's length, I saw the hummer losing strength.

I slowly reached out with my hand to catch what now could barely stand.

Through porous bones I felt its heart, its tiny, racing, living part.

With lightest touch, I held the life that on each front found danger rife.

I gently placed it in the sack, crawled 'round the rafter beam and back

to shaky ladder down I crept while female ruby-throated slept.

When finally down upon the Earth, I ope'd the pouch and gave rebirth.

It flew to food, red blossoms fair, and from garage, the man-made snare.

COWBIRD AND MINK

In a hole in a tree, on a twig in a nest,
cowbird laid a speckled egg after mother laid the rest.

The last to be produced, yet still the first to hatch,
it was biggest in the nest and the bully of the batch.

Always food for cowbird, yet rarely for her own,
the mother knew her duty, feed all nestlings until grown.

Brood now left in hardship, with offspring yet to fledge,
when predator came closer to a tree along the edge.

Up a tree, in a hole, mink cornered little meal;
now birds were unprotected and the mammal, poised to steal.

Pressing in still deeper, its space becoming tight,
the mink was quite immobile as the dawn broke morning light.

From a hole in a tree, on a twig in a nest,
we flung afield the body of each uninvited guest.

WYALUSING IN WINTER

An engineering friend of mine from third floor of our dorm
  consented once in early March before it got too warm
  to visiting the winterscape where I in Summers worked.
  I'd see the park as others did, official duties shirked.

Point Lookout hadn't moved an inch, its view remained the same.
  Still, we got the urge to run for something of acclaim.
  And there it soared, before our eyes, a hundred feet below,
  the brilliant white of eagle's head was scanning to and fro.

Its movements showed dynamics well, which pleased the engineer,
  but statics, now, I had to find, to make him really cheer.
  I led him to a "cave" of sorts, an icy overhang,
  where freezing water falls to Earth hits rock with liquid bang.

In Winter, liquid water pools and freezes on the floor.
The same when static ceiling-drips freeze downward more and more,
  until they reach the growing mound of ice in early Spring
  then form and melt the waterfalls that warming air will bring.

Engineering uses physics to study how moving things work together
(dynamics) and how non-moving things work together (statics).

THE COUNT

Before we paddled park canoes on weekly guided float,
I always talked (or did I boast?) of nesting birds we'd note.

One morn I told two twins we'd count, in case I often lied;
I wished to show my word was good, for me, a source of pride.

The siblings, both fine Catholics, knew all their prelates well,
including Protonotary, though difficult to spell;

Prothonotary Warbler, too, the quarry for our trip;
"We'll listen for its 'zweet, zweet' song," my one and only tip.

By knowing what their voice is like, "We'll count ten," I averred.
The girls were young and liked to learn, especially brand new word.

To let a little tension build, I warned, "We must not fail."
At my request, their count was mute till fully 'round the trail.

Their parents in the next canoe, the daughters were intense;
they wished to show to mom and dad adult intelligence

through bursting with their happy news of birds seen on the route.
"Our number came to twenty-nine!" (There never was a doubt.)

TREE FALL

I led our fleet that gifted day
along the river's shore.
Our canoes divided water,
my vessel in the fore.
As was my wont, I talked a while
before we paddled more.
'Tween islands lush we moved our boats
as many times before.
But this time came a thing so rare,
it added to my lore.
I knew not what would happen
nor what was for us in store.

This common, rarely-seen event
happed with no human hand:
the oldest tree before our eyes
fell over to the land.

INCREDIBLE EDIBLES

One year while working at the park, I did a cooking show.
The weekly program was a hit, once filled up every row.

With forty to one hundred guests, I had to move with haste
to serve each one in audience a little sip or taste.

The food I served I cooked that night or in preceding week.
Before the meal we'd take a hike, more edibles to seek.

One night a man at table last, refused to eat the nettles.
He'd seen all other groupings eat the nettles dish from kettles.

He'd even watched his family eat and say to him, "Great, Dad."
But still refused… Then ate a dab. He grudged to say, "Not bad."

Another night, when show was done, a couple came, was glad.
When you said nettles had the taste that spinach also had,
we worried that our son would balk, for spinach was a pain.
He loved the nettles best of all, we really can't explain.
I spoke to them, behind my hand, Here's your new spinach plan.
Next time there's spinach on his plate, say, "Nettles from the can."

MENU
Tossed salad of edible greens
Soup: Dandelion broth
Appetizer: Morel mushrooms (from freezer)
Entrée: Nettles Italiano
Dessert: Wild ginger candy or
             Berry flummery (by season)
Beverage: Nettle tea

(Adapted from *Billy Joe Tatum's Wild Foods Field Guide and Cookbook*)

175

SHELTER

With our country at its poorest,
some young men made it rich;

they built us timeless treasures while
in governmental hitch.

In Wyalusing, shelters rose
at hands of CCC.

Though built for humans, used by bats,
without an entrance fee.

Bats congregate 'neath roofing boards
and squeak like mice on high.

They face adapted predator
that doesn't even fly.

A black rat snake can use its scutes
as Caterpillar® tread

to climb the rough-hewn timbers where
on bats itself is fed.

WHO SAID I DIDN'T GO?

The air was fresh, the breezes light,
the grass had dotted dew.

The Sun began to warm the Earth,
the sky turned brilliant blue.

While hiking we saw many birds,
I named each one in turn.

The songs and calls of birds we saw
were next for them to learn.

I noted bill diversity
that matched the way they ate:

a cone-shaped cracker, sipping straw,
and chisel for a trait.

Two rivals hunted same resource,
each living by its tally.

The swallows caught bugs on the wing,
the pewees caught by sally.

With unlike feeding styles, the birds
stayed out of other's way.

They taught us how they share the wealth
while eating insect prey.

We saw more birds that blessed hike
and heard a Sunday crow.

When hikers talked of church, one said,
"Who said I didn't go?"

GRASSHOPPER SPARROW

My dad grew up on sandy farm in center of our state,
bucolic lessons in his heart, experiences great:

truck farming spuds and roofing barns with ice cream as a goal.
He'd beat the heat, once chores were done, down at the swimming hole.

His lessons in biology were making heifers cows,
'tween shearing sheep, he learned the birds who sang while sheep would browse.

It always seemed he knew the birds and mammals that we saw.
With mentor set, my goal was clear – know birds without a flaw.

One day he pointed in the sky, said "nighthawk" with a smile.
"How do you know? What makes it so?" He said, "I guess, its style."

Not learning from his farm-taught ways, I gave to them a pass.
Instead, I thought, while off at school, I'd take a birding class.

In May one year, while home from school, we visited a park.
I wished to show him birds I'd learned, like cuckoos and a lark.

When we saw jays and chickadees, my dad was in his prime,
but warblers, thrushes, vireos would stump him every time.

Most birds we saw were new to him, I didn't want to chide,
but finished with a bird ID that might have hurt his pride.

When something buzzed, my father said, relief upon his face,
"At least I know what made that sound – grasshopper in this place."

I knew the call and whirled around to see the birding treat,
Grasshopper *sparrow* on a stump, my triumph bittersweet.

MOTHERHOOD

En route one morn to birding place, where mighty river flowed,
saw I bare shell of hollow tree, collapsed upon the road.

While dragging off the bark-less hull, I saw a broken egg,
then two then three then more than four, then duck with injured leg.

The hen had found an egg intact and clutched it in her wing;
she shivered by the lifeless egg, but instinct made her cling.

I scooped the female wood duck, for she couldn't walk that far;
and placed her in a cardboard box in front seat of my car.

I ran the heat to keep her warm and drove her toward the shore,
next, waited for two passing trains, one stopped and went no more.

The train that stopped held railroad men, who checked the crossing gate.
Determined then to reach the shore, I seized from them her fate.

The bird, I carried o'er the tracks and set in sunny place.
She paused at length, near water's edge; concern spread 'cross my face:

the hope, not hers, but mine alone, for outcome that was good.
She bolted in, and claimed her chance at future motherhood.

TWO BIRDING CLASSES

I took a weekly birding course, two lectures and two labs.
Attendance most important, so professor kept close tabs

on outdoor labs with early start at February's close.
Once went to look by power plant, where not all water froze.

The wind chill fifty-odd below, we mostly just saw fog,
some mallards and black duck or two that shivered on a log.

Objective lenses of binocs and their eyepieces, too,
were frosted like my spectacles, which ruinates their view.

We found ourselves, just one month on, in warmer birding park,
the temp'rature was fifty then, *above* the zero mark.

Eleven species waterfowl plus other water birds.
I first saw rarest of them all, but didn't say the words

for I knew not the rarity white-fronted goose I saw
was for the teachers of the course, whose knowledge was not raw.

Goose Pond produced more birds that day, enough to whet our thirst
for seeing rare and migrant birds, like goose that I saw first.

A CHANCE ENCOUNTER

One morn while walking down a London street,
Bird-watching couple did I chance to meet.

They wished to catch and join a larger group,
with experts, we would follow on a loop.

I learned the European robin's song,
with perky notes, some buzzy, kind of long.

So many birds, I seemed a little boy
just learning magpies' sorrow and their joy.

A joy of friendship came to me that day
like artwork chalked upon the pavement grey.

I called them later, asked if they were free
next day or after for a cuppa (tea).

Mostly, I'm timid and not very brave,
Except if others would want Earth to save.

We dwellers of Earth, should take a long view
and give Earth all the respect that She's due.

KILLDEER

While jogging 'long the Oakwood Road, a rural city street
in the County of Milwaukee, where Nature, suburb meet,

I saw a bird just sitting there, in shoulder of the lane.
It sat until I got too near and triggered ruse to feign.

With open nest upon the ground, it made a daring ploy,
it limped away with 'broken' wing, behavior never coy.

It led me as a 'predator' with thoughts of easy meal
until the plover flew away, squawked "killdeer," then, with zeal.

Rejuvenated from its 'wound,' it ran in fits and starts,
apportioning its getaway in many shorter parts.

The next day, for my morning jog, I chose the other way
to give protective parent rest while at its nest could stay.

I looked not back, for I had seen behavior at its best,
to guard the eggs or nestlings that were camouflaged like nest.

I knew I'd miss the chance to see that bird repeat its trick,
but rather give best chance at life to each and every chick.

THE DUEL

So pleasant was the evening air and moonlit were the skies,
the crickets in the distance chirped, sweet fragrance on the rise.

The pipes were clogged, I washed by hand the night we had our duel.
I only wanted dishes clean and never to be cruel.

The wind picked up and rustled leaves, light rain began to fall,
an owl let out a warning hoot, and darkness cast a pall.

I washed the dishes, one by one, then put them in a rack,
next, took the dishpan through the house to empty from the back.

There was a sound, some feet away; I wondered what it was
because I know night noises not as well as Nature does.

I heaved the water anyway, but acted as a fool;
my toss became initial shot in very short-lived duel.

I saw no beast until it turned, then black back showed stripe white.
The skunk did not with rancor aim, yet squirted me with might.

Climbing TREES

When young, I used to climb the trees in our one-acre yard.
The larger trees had trunks mature, which made the climbing hard.
The rules I used for climbing trees is why I never fell.
The wisdom of these guiding words applied to life as well.
When lighting first begins to dim, go not too far out on a limb.
Use always three points of support: self, knowledge, God of any sort.

While on a trip
to trees Northwest,
where rainfall helps
them grow their best,
I left my climbing
days behind,
no longer feeling
so inclined.
Because the trees
were much too tall,
without the skills,
I'd likely fall.
No ropes, no spurs
were ever tried.
I climbed enough
to rest inside,
within the strength
of sturdy tree,
so safe, secure,
it let me be.

FARE WELL

The stream course through our piece of driftless land,
where we decided we would make our stand,

had few non-native plants with which to cope,
and fueled our thirteen-acre Eden hope.

Our lives were bliss with tree fort, cabin, bridge,
the only man-made structures near our ridge.

Until more yellow-flowered plant appeared,
we tried to keep our slice of Heaven cleared.

Invading force, non-native, just too strong,
head-start for occupation much too long.

Wild parsnip did not plan to overrun,
but grew and grew with energy from Sun.

Sun activates a chemical within
that causes sap of plant to burn the skin.

Just two of us to thwart the plant parade,
regretfully goodbye to land we bade.

BIRTHDAY PRESENTS

My wife enjoys her birthday date,
it always falls in March.

She never asks for birthday cake
with sweeteners or starch.

She likes instead to go outside,
clear weather, rain, or snow

to dapple in the sunny spots
along the river's flow.

Her greatest thrill is eagles seen,
in air, on ice, or nest,

or trolling from an ice floe chunk
that gives birds' wings a rest.

This burst of Life at Winter's close,
when eagles congregate,

one hundred thirty numbered once
and made her celebrate.

BALD EAGLE

Elegy and Ode

First, poisons, shootings, DDT,
    then habitats we didn't see.

How morbid was the missing news
    of eagles living on a lake.

How happy, though, the silly sound,
    the chuckle call of eagles make.

When Earth's returned to Nature's state,
    the eagles miss endangered fate.

Stories

Perspectives of Nature Collected Works

Index

SW=Selected Works,
CW=Collected Works

About the Author

The scientifically romantic nature poetry of Paul Košir has its academic roots in his nine years as a student at the University of Wisconsin-Madison. There he earned bachelor's degrees in math, natural science, and history. In 2010 he received a master's degree in natural resources and environmental education from UW-Stevens Point.

The experiential poetry was drawn mostly from his twelve years as the naturalist at Wyalusing State Park near Prairie du Chien, Wisconsin. He also drew on this background to write articles for *Wisconsin Natural Resources* and *La Crosse Magazine* and to publish the book, Wyalusing History.

Košir has taught biology, physical science, and math at the high school level and earth science, biology, and environmental issues at the college level. As a naturalist, he taught all ages about nature through hikes, programs, and displays, something he still does occasionally as a volunteer.

He was also privileged to lead hikes for cognitively and physically challenged individuals, as well as worked as support staff and teacher for that population.

Born in Milwaukee, Košir now lives in La Crosse (soon relocating back home near Milwaukee) with his wife and their two sons. He enjoys writing, hiking, bird-watching, gardening, traveling, and visiting relatives.

Paul Kosir

Also by this author:

Wyalusing History

Perspectives of Nature

Perspectives of Nature Volume 2

Perspectives of Nature Volume 3

Perspectives of Nature Volume 4

Perspectives of Nature Selected Works